# The Rogue Perfumer

Dr. Bobbie Kelley

Library of Congress Control Number: 2009903746

ISBN:
Softcover 978-1-4415-2909-1
Hardcover 978-1-4415-2910-7



This book was printed in the United States of America.

To order additional copies of this book, contact:
Xlibris Corporation
1-888-795-4274
www.Xlibris.com
Orders@Xlibris.com

If your blood were perfume,

Where would I keep it?

In the bedroom?

Would I sell it on the market

For the other girls to wear?

Or keep your sweet blood to myself?

Or do you really care? . . .

To know,

To feel,

To smell your sweet perfume . . .

***Bobbie Kelley***

*1985, Gretna, Louisiana*

Contents

# PREFACE

What is it that inspires one to create a perfume? It can be love, sex, money, a stroll in the park, animals, passion, desire, happiness, past lives, swimming, travel—in fact, anything. Just as it is primordial to smell, so it is to create a perfume. As long as there are human beings on the planet, you will have perfume. The Webster dictionary, tenth edition, defines perfume as "the scent of something sweet-smelling; a substance that emits a pleasant odor, esp. a fluid preparation of natural essences (as from plants or animals or synthetics and a fixative used for scenting); to fill or imbue with an odor."

I have worked with thousands of perfume materials, both what is labeled *all-natural* and synthetics, and every raw material has its place in the making of perfumes except for the ones that I have listed in the book as being harmful to an animal's life. I have also made hydrosols, face creams, lip glosses, shaving cream, sugar scrubs, massage oils, etc., and I always go back to perfumes.

A person could study every known history book in the Library of Congress on the subject of odor, and every book will eventually come to some estimate on the date and place that perfume started, but I will tell you that perfume is older than even that. It is a basic human instinct to smell and to smell others. It is primary to survival both in humans and animals, and we would not be on this planet without that sense. Science is still discovering about the pheromones that humans emit that make them attracted to one another. The animal kingdom is full of stories on how animals meet, mate, and survive by their sense of smell. But this book is not on the history of perfumes, or about animals; rather, it is about the art of perfume creation. Great perfumes are truly magical, so beautiful as to seem supernatural, for they have their own persona created by their maker.

I will be disclosing a lot of my personal thoughts on this subject, and I will write freely to expose my feelings and, especially, my soul as I am writing, so I will not be holding anything back. Therefore, please keep an open mind as you read this book, and I do hope that you can find some humor as you embark on your perfume-creating journey.

To be a perfumer, you don't have to be a chemist. You don't need to have anyone's permission, you don't have to move to Grasse, you don't have to have a famous grandfather who was a famous perfumer from France, you don't have to know someone or be a duchess or baron, and for that matter, you don't even need a college degree. Just be you, study hard, be creative, don't hold back, be passionate, possess lots of money, and, especially, have fun with all of this. And always remember, the planet is your perfume playground!

The following abbreviations that I will be using are more or less universal, so it is helpful to know and understand these. And some I have made up myself, and they are my own language.

| | |
|---|---|
| With | w/ or /c |
| Without | w/o or /s |
| PEA | Phenyl ethyl alcohol |
| PEACE | Phenyl ethyl acetate |
| Ben | Benzoin |
| Frank | Frankincense (a.k.a. incense) |
| Ant. Methyl | Methyl anthranilate |
| Abs. | Absolute |
| HDXC | Hydroxycitronellal |
| Gm or gms | Gram or grams |
| Oz. | Wizard of Oz—just kidding, Ounces |
| Ald. | Aldehyde |
| Fix | Fixatives |
| Y/Y | Ylang, or ylang-ylang |
| Jaz | Jasmine |
| MNK | Methyl napthyl ketone |
| Clary | Clary sage |
| Tinc | Tincture |
| Syn. | Synthetic |
| A.k.a. | Also known as |
| Incense | Frankincense |
| Ml | Milliliter |
| CP | Compound |
| BOB | Bobbie's Oriental Base |
| Sniffer | Nose |
| Patch | Patchouli |
| LOL | Lots of laughs |

I will also be assuming that you know most of the raw ingredients for perfume making; however, if there is some ingredient that is rare or for any other reason unfamiliar, I will elaborate on it. I will also tell you where I got the ingredient and any other information about it that might be relevant to the perfume formula. This book is not necessarily for a beginner perfumer, although a beginner might find this most helpful in their learning process. It would be to their advantage to research the Internet and/or read the reference material that I suggest.

Also, I will not be restricting any ingredients or speaking of restrictions on the raw materials. As I said before in my intro, I do not appreciate someone telling me what I can and cannot put in my formulas. Usually, the ingredients that have hurt people were extremely rare, and there is usually something else involved. Besides, I can think of a million other things that would be harmful to, or even kill, a human. Most perfume ingredients are safe, and I have never known anyone to die from wearing a perfume that is made by a professional. Most humans who say they are allergic to perfume have never even been tested, and I believe most are reminded of or remembering a bad experience related to a smell. This will usually always come to clear light during a hypnosis session. They aren't even conscious or aware

of what it is exactly that is making them feel that way, so they blame perfume. Furthermore, people also say they are allergic to iodine, peanuts, shellfish, chocolate, feathers, down, etc., and these certainly aren't outlawed.
I am also well aware of the controversies revolving around the use of synthetics in animals. The information written on material safety data sheets record testing lethal doses in animals which is absurd. Even an ingredient that we consider "all-natural" and organic can also become lethal at the same dose given to animals or humans. Furthermore, the lethal doses that are injected into animals are astronomical, and there is never that amount placed in a perfume. If anything, one could possibly be receiving a homeopathic dose of the material in question. Bottom line is this: we are not animals, and I certainly don't support the testing of anything on animals unless it is specifically for animals. I wish more humans would step up to the plate and, if they are so concerned about this issue, have the research done on themselves and not on the animals. I am more than willing to do this. Animals should not be forced to wear perfume or cosmetics or any substance in them.

There should be warnings on packages of perfumes, just like in cigarettes or any other product with warnings and contraindications, saying what could happen to you if you were to wear this perfume. It is not as if perfumers are sitting at their organs working with raw materials such as toxic mushrooms and poisonous hemlock; leave that to the huge drug corporations. My point is this: restricting a perfumer's ingredients really stifles his/her creation. I used to work for the FDA testing and researching drugs, and I can tell you that just by taking Tylenol or ibuprofen, you can destroy your liver, so please, spare me the bull—. And the day they changed my notes to save a few million dollars on a research project when I spared a man's life is the day I said good-bye to Western medicine.

I do, however, use terpeneless oils as much as possible, not because of sun sensitivity, but because of oxidation and rancidity. As far as light sensitivity goes in regard to terpenes, well, I say, get out of the sun; it supposedly causes cancer and makes your skin look like an alligator's.

Also, I would like to add that I am very fortunate that I own my business and that I am able to disclose these perfume formulas to you. If I were working for a big, or even a small, perfume company, I would not be allowed to share this proprietary information with you; nor would I even if I did know their secret perfume formula. The formulas, besides my own, have already been published, and I give credit where it is due. It was my personal decision to share my formulas with you. Also, I would like to add that I have not been paid or bribed in any way by a chemical company mentioned for advertising their raw materials.

I found it very challenging to write this book because of all of the information regarding the making of perfumes, and it was hard not to make this into an encyclopedia. I really wanted to stay focused on perfume formulas and their creation and not on the history and who said what, when, and where. There are already too many books on the history of perfumes and not enough published formulas.

I admit that I am not a good writer, and since this is my first book, I hope that you will forgive me as I stumble through this writing journey trying to transmit to you my formulas, thoughts, lessons, and, most especially, my feelings, which has always been my most difficult task throughout my life.

If you are a beginner perfumer and you are reading this book, I want to say, don't ever let anyone stifle you or your creativity while you are at your table/organ. And by all means, make mistakes; they are our largest lessons. Just be sure and document everything, even if it makes you feel silly or stupid. Do not ever put yourself down or your creations. Take your play home with you. "Feel" your creations and not just "think" them. Smell everything!

# ACKNOWLEDGMENTS

I want to thank the following people and animals for helping me with this book whether it was through inspiration, unconditional love, support, courage, knowledge, or physically helping me edit. I don't know if I will ever be able to thank them enough, so I figure I will forever be in their debt. Without them, this book would have never happened, and I am so grateful.

God
My Mom and my Dad, Mr. and Mrs. Henry and Julie Kelley
Penny Kelley, who first reminded me that all things are possible.
George Welman
Jerry and Esther Hicks, who also reminded me that all things are possible.
Ken Hertz
Dr. Leo Murakami
Xlibris and all of their staff
Ernist Nelson
Juan and Cindy Dawson
Jack Condy, who is my ninety-five-year-old friend
My dog Honey Girl, who keeps reminding me that all things are possible.
My other dog Tango
My English teachers
Rene-Maurice Gattefosse, Master Perfumer, who never forgot that all things are possible.
Arcadi Boix Camps
Danute Anonis
Roman Kaiser
Editor Judith Cruz
Editor Liz Calledo
Lori Adams, Author Services Representative for Xlibris Book Publishing Company
John Signe, Market Consultant for Xlibris Book Publishing Company
Andrea Reeves, Submission Representative for Xlibris Book Publishing Company
CB Coleman, Sales Representative for Xlibris Book Publishing Company

I want to also thank and acknowledge the following companies:

Vigon International
Ferminich
Bio-Botanica
Bedoukian
Pearl Chem.
Mane
Citrus & Allied
Givaudan
Quest
IFF
Millennium
Lipo Chemicals
Mitsubishi Gas
Frutarom
Liquid Alchemy Labs
Remet Corp.
Ungerer & Company
*Perfumer and Flavorist* magazine
Allured Publishing Corp.
Fuerst Day Lawson
Xlibris Book Publishing Company

I would also like to especially thank all of the chemists who are working so hard on raw materials.

Aloha, and thank you to all of the other perfumers and flavorists of the world!

***Bobbie Kelley***

# INTRODUCTION TO THE BOOK

I was propelled to write this book originally to share some of my proprietary perfume formulas with you, which are very rare in the perfume world. And if you do nothing but just study these formulas, you will have gained an incredible amount of useful information that you will not find anywhere else in the world. Perfume companies sell their formulas for thousands, if not millions, of dollars; and when you purchase a perfume company, along come their formulas out of their vaults that are included in the intellectual property, and so these formulas come out of mine. I can guarantee that the price you have paid for this book is insignificant compared to the knowledge you will have gained from this information. It has been said that knowledge is power and I have to agree with this statement. After reading this book, you will have the information to make perfume, or at least be inspired to make a great perfume, and who knows, perhaps even a better one than before.

I have also added my opinions, excluding so-called facts, because in the world of perfumery, there is not so much of what we humans term fact. Nothing in perfumery is factual, and a beginner perfumer would do well to remember this. If you are a professional perfumer, I invite you to leave the world of fact behind you as it can distort your creative process; step outside of yourself and go beyond what is so-called factual or reasonable. These are the qualities of a great perfumer, a person who is bold and courageous and who does not accept things as they are, but what they perceive them to be.

The world does not ever get a chance to really know about perfumes or the perfumer, but just the mystery of them, and so this book allows you to step inside the clandestine vault and the mind of the perfumer.

# INTRODUCTION TO THE AUTHOR

I would like to tell you a little about my background because I think it is important and relevant to this book. Firstly, I was born in Kentucky, and I am of Scottish descent. My grandfather had a well-hidden still in the hills of Kentucky, and not only did he make moonshine, he drank it as well and shared it with anyone in the community who might be ill. We always had a bottle of moonshine in our home for when one of us might become ill, and it always helped when we were. So I guess you could say that distillation is in my blood, and so are the wonderful smells relating to the making of alcohol: the herbs, flowers, and the beautiful woods in Kentucky. My grandfather also raised hemp for the federal government, and not only did he smoke it, he always shared it with the agents that came from the main headquarters to check on the crop. He was of a very high social status, was very much respected by the community, had a huge heart, and had a great sense of humor.

Both he and my father taught me a wealth of information about the land, animals, plants, trees, and, especially, tobacco. Since Kentucky is one of the most planted grounds of tobacco in the United States, it was natural that I was to learn about it. It is also now one of my favorite perfume ingredients. I remember having an earache when I was young, and there was not a doctor in the county that could cure it, so my father blew tobacco smoke in my ear one night as a last resort. By morning, it had completely healed.

Looking back, I realize now that I was born to be a perfumer. I smelled everything in sight and was always on the lookout for new and different smells. I hungered for them. I still do.

I can remember riding in the truck with my father, and when he stopped to pump gas, I would always roll down the window just to smell the petroleum. I would ask him a million questions about where it came from and why it smelled like that. In hindsight, I must have driven him a little crazy constantly interrogating him, as if he would know all the answers, about why certain things smelled the way they did. Everything was intriguing to me, not so much because it was the object but because it had an odor. Everything still is.

Then into the kitchen with my mother, the same scenario would take place. Where does cinnamon come from, and why does it smell that way? I would ask. Well, what about cumin, paprika, pepper, basil, tarragon, rosemary, thyme, and on and on . . . Schooltime was one of my great loves—the smell of the new books, my cedar pencils, my cedar pencil box, etc. During class, I would love to go to the pencil sharpener just to smell the cedar shavings. I remember having forty-seven pet mice when I was a kid, and I used to love to clean their cage because I loved the smell of the cedar shavings. Sometimes I really think that I did so well at school because of the odors; they somehow stimulated my brain into learning.

I never played with dolls, but I was always fascinated with the way the plastic smelled, or even my Tonka toy trucks, their tires; my plastic army men; my play guns and their toy holsters; my PF flyers; my softball glove, anything made of leather; the flowers, trees, dirt, bugs, grass, glue, especially airplane model glue; etc. I always joke and say that it is a good thing my parents never bought me model airplanes to put together because I probably would have loved sniffing the glue! Fact is, I never got tired of smelling things, and I still don't. It is hard for me to comprehend what is called nose fatigue. I never tire from sniffing things. And I do not believe the reports that when people get older, they lose their sense of smell; I say "Poppycock!" on that. That is ludicrous! I have kept my sense of smell alive. As a matter of fact, the older I get, the sharper my sense of smell becomes. Is this training? Is this because I believe it and therefore make it true? Is it because I am different? Is it because I take care of myself? Is it because I came from outer space? Well, I guess it could be any of the above, but the point I am trying to make is that we should never believe things just because someone said so, and that certainly does not make them true. What I do get tired of is hearing all of the negative things about the human body and what we cannot do. I can tell you that I have been at my perfume table for hours on end without ever having to sniff coffee beans or put a piece of wool in front of my face.

Then it was on to nursing school in Baton Rouge, Louisiana, where I was almost kicked out for apparently "practicing voodoo" (lots of laughs); and yes, this was still the 1980s. Not only am I a clairvoyant and medical intuitive, meaning being able to look at a human body and see disease, but I am also an herbalist. This did not go over very well with the nursing instructors when word got around that a fellow student, also from Kentucky, and I were going to New Orleans every weekend. New Orleans is known not only for its decadency, which is one of the things I like about it, but also for its herbalists and psychics. I, of course, had also been seen talking to "colored people," and word got around that I had them as friends. I was really in trouble then. So I was called into a little room and was interrogated by five Nurse Ratcheds, and I was to place my hand on the holiest of the holy, the Bible, and swear that I would never practice my "voodoo" at the hospitals ever again or tell anyone that this conversation had ever taken place. They had said that it was "the work of the devil." I did as I was told because I wanted very much to graduate, but I also told them that I would still talk to "colored people," and that they would always be my friends. In hindsight, I really wish I had told them all to go to hell.

I must add that I was in awe of the smells related to the medical field: the iodine, ammonia capsules, cold metal, and disinfectants, probably due to the pine, another one of my great loves. I did learn the metric system (something not taught at the time in public schools), pharmacy, the law relating to medicine, and, most importantly, how to take meticulous notes. This has helped me tremendously in the world of perfumery. I also studied alcohols, including absinthe, in detail and learned the art of professional bartending. Little did I realize at the time that this would also help me tremendously in the world of perfumery.

Then later, it was on to Chinese medical school, where using your psychic intuition is always acceptable and acknowledged, actually encouraged; and for that, I am grateful. I was able to sharpen my herbal skills making tinctures, macerations, infusions, and concoctions of all kinds. The only trouble I got into there was from a teacher; I am still wondering if she was the daughter of one of the Nurse Ratcheds, and she taught an ethics class in which I wrote my thesis on "The Art of Sexual Healing." That did not settle with her very well, and she tried to have me banished from the school. Not only did all of the other teachers love the writing, they all defended me and advised the ethics teacher to have therapy; and at that, she immediately resigned. While in medical school, I also became a chef and, at the time, did not realize how this would help me in perfume making.

From there, I was led to hypnosis and became a clinical hypnotherapist and master hypnotist. The main reason being I kept having "olfactory visions" of me being a small baby in the backseat of a car on a woman's lap, which I felt was my mother's, and smelling her glorious perfume. In the vision, I remembered seeing the Statue of Liberty out of the window. When I asked my mother (in this life) about ever being in New York when I was a baby, she laughed at me and said that I had never been there before, nor have I been as an adult. So I needed to do some past-life regression therapy to find that beautiful perfume that the woman was wearing in my olfactory vision, even if I had to make it myself. The perfume was exquisite. Is there really such a thing as past-lives, and if so, is it possible that we bring our sense of smell with us into our present lives? I believe that we do. I have taken people on amazing "sensual" and olfactory journeys during hypnosis and witnessed incredible scenes where one can remember, in minute detail, an entire life, led by their sense of smell.

So, knowing a little of my background and from where I hail, you should realize that I am not the kind of person who follows a lot of rules or is brainwashed into what other people think might be good for me, especially the federal government. I don't like to be told that something is illegal because taxes cannot be collected on it. I am honest, genuine, and I say what is on my mind. I am not fake or pretentious. I do what I think is best for myself and for those around me. I will decide what I apply to my skin, and if it kills me then that is my free choice. I am a rebel, I am a nonconformist, and I still question everything. Even though I was born in the United States, I do not claim to be an "American," and I especially do not support war of any kind for any reason. And if I were in your country, I would not follow your rules either. I guess you could say that I am a woman without a country. I am a rogue perfumer.

# CHAPTER 1

# Leather / Russian Leather Perfumes / Animal

Perfumers are different, and in saying that, I have to be very different. I love the family of leathers, and it is one of my favorite families of perfumes. Some of the earliest recorded perfumers in France were perfumers of leather. As far as leathery odors go, I have never met a person who said, “Oh, the smell of saddles and baseball gloves really turn me off,” or “I can't stand that new smell of leather in my brand-new Mercedes.” And if you ask most women and men about the number-one attraction of their beloved while under hypnosis, besides their spiritual energy, they will tell you that it is the way they smell. Most people are not even aware that it is their partner's odor that they are attracted to, most probably due to unawareness and lack of sensuality. Just as in music, there is such a thing as overtones, which most cannot hear, but the notes are out there in the ether, nonetheless. This is the concept of pheromones, and as science will soon discover, it is my belief that there are many more. I also believe a human can be trained to sense or smell these human odors or those we could have had the ability to smell centuries ago and lost for various reasons.

Have you ever met a person that smelled so good that you wanted to bottle them? This is a very natural and a typical example of sensuality. My father smelled so good that all of my five sisters would fight to sit on his lap because we wanted to know his secret of why he smelled that way, and he was a smoker! My mother swore and pleaded for him to take a weekly bath, but only because she believed more in cleanliness. When I was up to some mischief, I always knew my father was home from work, not because I heard him walking up to the door, but because I could smell him. There are times when I am working at my perfume table that I can still smell him. I have heard of this phenomenon in the company of others who have studied the paranormal and that say that when they sense, feel, or see a ghost, it is usually preceded by the way they smell.

So I have wanted so desperately to find the magic in my father's smell, which makes me think and chuckle over the wild book, now a movie, *Perfume,* by Patrick Süskind. As morbid as it sounds, what perfumer has not thought of bottling a human odor? No, I did not kill my father, nor have I ever thought about actually murdering someone for their scent, but my father's odor is somehow implanted in my olfactory pathways, and I will take it to my grave. By the way, I did so enjoy both the book and the movie.

Perfumers must be open to all smells, this being much easier said than done of course. I believe that the foulest odors are the magic in great perfumes. I will speak more about them in this book. I will also include some classical formulas for perfuming leathers.

Leather-type odors can include labdanum, costus, cade, ambrettolide, grisambrol, all musks, exaltolide, amber, civet, castoreum, ambergris, styrax, skatole, tobacco, cresol, creosol, woods, birch tar, various phenols also used for preserving leather, etc. There is even synthetic leather, but I do not recommend using it.

Here is a little story that I think of when I think of the ingredient costus, which, by the way, is very expensive and on the dreaded restricted list. So why use it? It is one of my favorite ingredients, and it has been said that it smells like a sweaty men's locker room or dirty socks. I was working at a nightclub while I was in medical school in Santa Fe, New Mexico, in the early '90s, and every Friday night (payday), a strange man would come in and ask to pay any girl $100 for their pair of dirty, stinky socks. They had to be very well worn and very grungy. It became so much fun for us girls to see how long we could wear a pair of socks and then, of course, let them marinate in our lockers. We even had a locker designated just for the "stinky-sock guy." Occasionally, one of the socks would disappear like in your dryer and would fall back behind the huge wall of lockers, so we could only get $50 for one. It was a lot of fun, and I have fond memories of that time. The man never disclosed any information as to why he was doing this, nor did we ever know his name or anything about him. None of us would ever ask because it became so much fun to make up stories in the locker room and try and guess. He was very mysterious and intriguing, and I suppose all of us girls really liked that mystery.

I had an old artist friend who once told me one time that when he was in college, he loved to boil old leather shoes and smell them while he was studying. I always thought that was strange, but hey, whatever works. He was a genius, and one of the most amazing artists that I have ever met.

Back to leather perfumes, this chapter would not be complete without mentioning the very classic and classy Russian leather made by Ernest Beaux, Master Perfumer, in 1920, which is a perfume masterpiece. One whiff of this magical potion could send any man or woman into an ecstatic state of having multiple orgasms. No, just kidding, but I thought I would say that in case you are falling asleep. I am fortunate to have an original two-ounce bottle of Cuir de Russie that I paid US $800 for in 2006. Besides Ernest being Russian and having a background of curing leathers, he was a perfumer genius. I imagine that Ernest must have had a red-hot lover with an incredible smell that he was trying to capture. One reason, I believe, that his perfume was so incredible was that he used a lot of natural ingredients; even though this perfume was marketed in France, the main ingredients probably came from Russia. Most of the original great perfumes were of local and natural origin, and I predict the great perfumes of the future will be made that way again.

Historically and up to the present, the Arabs, Egyptians, Greeks, Turks, Mongolians, etc., would urinate on leather to help loosen the hair and make it easier to clean. Camel's urine is still used in the Middle East. I believe that the thought of using civet in a perfumed formula came from the idea of using urine to cure leather. I was very fortunate to see and spend some time with a civet cat, which is not really a cat, in Singapore, and I was happy that they were not taking civet from her. Although there is probably nothing better than to make your perfume creation, it is also a very cruel and painful process to collect civet from the cat, so probably it is best to use the synthetic civet, which smells just as bad. I use synthetic civet in just about every perfume formula that I make. I do not ever use real civet in any of my perfume formulas, nor do I purchase perfumes that contain authentic civet or civet absolute.

There were many reasons to cure leather besides making it supple; one was to make it smell good. There is nothing worse than the smell of rotting flesh, especially the flesh of a human. The perfumed leather gloves that were the rage in Europe after the dark ages were not only soaked in civet, but were probably popular because they were supple and felt so nice to the rough-skinned hands from the men doing hard labor. There is a certain eroticism, as well as sensuality, in being touched by someone with not only exquisitely soft skin but skin that is perfumed. The women also had to have them, and these were their rage as well.

So here is a classic formula for perfumed leather, also known as Spanish skins.

**Spanish Skins**

|  |  |
|---|---|
| Rectified oil of birch tar | 5 |
| Sandalwood (authentic) | 25 |
| Bergamot (authentic) | 20 |
| Petitgrain | 20 |
| Lavender | 10 |
| Clary | 5 |
| Coumarin | 4 |
| Musk ambrette (authentic) | 10 |
| Rose (authentic) | 10 |
| Alcohol (your choice) | 900 |

The above formula is from W. A. Poucher, *The Raw Materials of Perfumery Volume 1*, also known as the Perfumery Bible, first published in 1923. A classic book, good luck finding a copy, as I believe they are out of print. At least they were the last time I checked.

I get a real charge out of hearing the term *Spanish skins*; my mind runs wild thinking of the Spaniards introducing this to the French, which is supposedly how the French learned. A Spaniard storms onto rue du Dragon, Paris 6th, on his beautiful stallion, which is probably an Arabian that he bargained for, for days in return for his beautiful Spanish sister, who became the Arab's fourth wife. Cloaked in his awesome flaming red bolero, having donned red leather perfumed gauntlets all the way up to his armpits, matching his red leather boots with a set of shining spurs, clicking his castanets while keeping time with the stamping of his horse's hoofs, he shouts in his deep, guttural Spanish accent, "What gentleman would like to possess my secret to having all the pretty girls begging for your affections?" In all likelihood, he probably learned it from the Arab as he was bargaining his sister for that beautiful Arabian stallion.

When I was traveling and studying in Egypt, there was not one Arab man that would not swear to Allah with his hand on the Koran that they are the true discoverers and original teachers of perfumery. They believe this with all of their Arabian hearts, so who am I to argue? After all, "his-story" is exactly that, "his story," so I don't place too much credence in it—meaning I would like to hear "her story" as well. But what I do believe is that the ancient Egyptians, who I believe do not exist anymore, did discover the art of curing leathers, that is to say, human skin, the process also known today as embalming. It is interesting to note that most of the authentic and so-called all-natural raw materials that were used for mummification are still used for perfumes today, and modern embalming is now done with formaldehyde.

I have a watch that is very well made, with a wonderful leather band, that I wear every day, and I occasionally use it to tell time. But I really wear it for is its perfumed odor. I love it, and I know that watches are now considered a little old-fashioned because people use their cell phones to tell the time, but it is impregnated with some of my favorite perfumes that I have made, including Russian leathers, which I wear on my wrist. I am well aware of the flavored and perfumed jewelry, plastic watchbands, and cell phones, but it is not the same. My watchband has the smell of warm resins, balsams, and real authentic flower absolutes.

Here is another formula for a Spanish skin from the book by George W. Askinson.

**Spanish Skins**

|  |  |
|---|---|
| Benzoin | ½ lb. |
| Oil of bergamot | ¾ oz. |
| Oil of lemon | ¾ oz. |
| Oil of lemongrass | ¾ oz. |
| Oil of lavender | ¾ oz. |
| Oil of nutmeg | 150 grains |
| Oil of clove | 150 grains |
| Oil of neroli | 1 ½ oz. |
| Oil of rose | 1 ½ oz. |
| Oil of santal | 1 ½ oz. |
| Tincture of tonka | ¾ oz. |
| Oil of cinnamon | 150 grains |
| Alcohol | 1 quart |

The musk and civet are added later in a different process, but are nonetheless in the formula.

The following are two classical formulas that are well-known Russian colognes:

**Russian Cologne by Felix Cola**

|  |  |
|---|---|
| Bergamot | 260 |
| Lemon | 280 |
| Lavender | 60 |
| Linalyl acetate | 30 |
| Rosemary | 40 |
| Alpha ionone | 100 |
| Neroli petals | 20 |
| Isoeugenol | 50 |
| Musk ketone | 30 |
| Musk ambrette | 20 |
| Coumarin | 10 |
| Vanilla | 40 |
| Heliotropin | 60 |

This formula dries out very powdery due to the last three ingredients, and I do not want a leather to smell powdery. If one were to have more of a leathery note, you would add civet, but since this is just a cologne, that is acceptable. Also, the rosemary tells me right away that this is probably cologne. If you would take out lemon, it would smell exactly like Creed's Russian leather, which is very boring. It actually should not even be called Russian leather and is very misleading. I usually like Creed, so I am very disappointed in their marketing of this. Recently, I purchased a bottle of Russian cologne from Russia and one from Germany, and they are exactly like this formula, so lots of people have copied this. Another brand of so-called Russian leather is Demeter, and should not even be called this for it disgraces true Russian leather.

I have been fortunate, however, to own some nice Russian leathers from the '60s that are actually in good condition, and that were made in California. None were made by a major fragrance company and, to my knowledge, do not exist today. Some of the ones that were nice were Imperial del Oro, Royal Argenta, Prince Obelinski, Aristocrat, and Saxony. These leathers fetch very high prices on the Internet and are extremely rare, so if you are lucky enough to get your hands on one of these, grab it fast.

**Russian Cologne by Mann and Winter (Austrian perfumers)**

|  |  |
| --- | --- |
| Bergamot | 40 |
| Lemon oil | 20 |
| Essence of Portugal | 20 |
| Rosemary | 4 |
| Lavender oil | 6 |
| Tangerine oil | 2 |
| Clove | 2 |
| Neroli (synthetic) | 12 |
| Methyl ionone | 0–5 |
| Oakmoss | 0–5 |
| Vanillin tinc. | 150 |
| Tolu balsam tinc. | 250 |
| Castoreum tinc. | 35 |
| Musk ambrette | 20 |
| Reconstituted amber | 10 |
| Ambrene | 5 |

This is, in my opinion, a much better formula than Felix Cola's. Be very careful not to use tinctures with a high percentage of raw material. For instance, the tolu balsam would have to be below 50% tincture; otherwise, it is too much and overwhelms the citrus notes that are so important for cologne. Also, this formula does not dry out as powdery as Felix Cola's formula; but of course, there is not any coumarin or heliotropin in this one.

In a formula from the antique book by Dr. John H. Snively, from 1877, Dr. Snively gives only three ingredients for a Russian leather that I find interesting, and they are patchouli extract, rose spirit, and European birch otto. I was lucky enough to stumble upon this wonderful original book on eBay, and I cherish it. It is surprisingly in good condition considering its age.

Famous perfumes with leather notes include the following:

Bandit (also considered a chypre)
Cabochard
Cuir de Russie
Miss Dior
Parure
Scandale
Tabac Blonde

Here is one of my Russian leather formulas:

**Russian Leather by Bobbie Kelley**

|  |  |
| --- | --- |
| Rose, synthetic | 70 |
| Texas cedar | 50 |
| Clary | 10 |
| Bergamot | 50 |
| Cinalkex | 10 |
| Jasmine, synthetic | 50 |
| Clove | 20 |
| Dihydroeugenol | 50 |
| Lilac | 20 |
| Gerionol | 70 |
| Grisambrol | 30 |
| Citronellol | 40 |
| Vanilla compound | 40 |
| Amber | 120 |
| Labdanum | 13 |
| Labdanum absolute | 10 |
| Exaltolide | 40 |
| Castoreum tinc. | 10 |
| Civet tinc. | 10 |
| Fixateur 505 | 50 |
| Tobacarol | 80 |
| Chypre base | 120 |
| Birch | 10 |
| Cananga | 80 |
| Styrax | 80 |
| Pine needle | 10 |
| Birch tar oil | 5 |

As you can add, most of my formulas do not equal 1,000, and it is very rare that they do. It is my belief that some formulas in books that do add up to 1,000 are usually missing something. I do know that in the food business, most chefs who divulge their recipes usually leave out a secret ingredient that you must figure out on your own. I believe that most perfumers do this as well. I, however, will not be leaving anything out, trying to make you guess; nor will I be trying to have it equal 1,000 just for the sake of a pretty formula. It is important for a perfumer to think out of the box and not get stuck trying to follow the rules.

To sum up Russian leathers, the two ingredients that are imperative to true Russian leather, besides synthetic civet, would be birch tar oil and synthetic castoreum. The company Vigon International has a wonderful synthetic castoreum, and although it is a little pricey, it is well worth saving the animal's life. Also, the team at Vigon is great to work with, and they are very fast, as well as professional. They used to carry Russian leather, but apparently, it is no longer available. I do not use real, authentic castoreum in any of my perfumes.

And thanks to Arcadi Boix Camps, God of synthetics, Master Perfumer, for all of the wonderful information in his book, *Perfumery: Techniques in Evolution*, published by Allured Publishing in 2000. He reminds us of the many raw materials that can be used in leather types of perfumes such as the following:

Dihydroactinidolide
Dihydroambrinol
3-Methyl-6-p-tertiary butyl phenol
Shangralide
Cyclohexadecanone or musk CHD
Musk ether
1-Oxaspriro (4,7) dodecane or Oxyvet
Nolinac
Muscacide
Muscambrol 850
Costaulon
Habanolide
Muscenone
Exaltonene
Novalide
Muscalone
Ambrox
Grisalva
Labdacore
P-Tertiary butylquinoline
Narcisse ketone
Sumatril p-Cresyl crotonate
Citrindol
Aldehyde NU and alcohol NU
Cashmeran
Celestolide

I had a lover one time tell me that my sexual organ smelled like a bed of roses, and that has always stuck in my mind. So I decided to take that idea along with the idea of vaginal pheromones and make a perfume with this concept in mind. Again, inspiration can come from anywhere; allow it to flow. This next concoction is an example of making a perfume in stages so I will use it to show how I do this. See more examples in Chapter six.

I took Secondini's original perfume formula of Ambergris Perfume No. 1 and played around with it until I got what I liked. This has an overall musky odor and was made in an archaic fashion using the old apothecary method of measuring; it just had to be done that way.

**Perfumed Vaginal Juices (a.k.a. Animalistic) by Bobbie Kelley**

Ambergris tinc. (syn. at 50%) .......... 50 ml
Ambrettolide .......... 10 ml
Rose (syn) .......... 2 ml
Vanilla tinc. (25%) .......... 1 ml

Five months later:

Ambergris tinc. (syn at 50%) .......... 2 oz.
Musk tinc. syn. ( 10%) .......... 1 oz.
Civet tinc. (syn. at 1%) .......... ½ oz.
Castoreum tinc. (syn at 10%) .......... 1 ml
Labdanum tinc. (10%) .......... ½ oz.
Jasmine, syn. .......... 2 oz.
Neroli, syn. .......... 1 oz.
Indole .......... 1 minum
Indolene 50 in DP w/ HDXC .......... 1 ml
Styrax tinc. (perfume grade 20%) .......... 1 oz.

Thirty days later:

| | |
|---|---|
| Grisambrol (Ferminich) | ½ oz. |
| PEA | 1 oz. |
| Exaltolide | 1 oz. |
| Cistus tinc. (10%) | ½ oz. |
| Skatole tinc. (1%) | ½ oz. |
| Styrax resin, purified perfume grade | 1 oz. |
| Hedione | ½ oz. |
| Bergamot, syn. | 1 oz. |
| Ambergris tincture (50%) | 50 ml |
| Ambrettolide | 10 ml |
| Rose, synthetic | 2 ml |
| Vanilla tinc. (25%) | 30 ml |

One year later:

I like it! I would, however, like to be very decadent and remake it using all-natural materials without, of course, hurting any animals.

As far as bodily fluids go, here is what Dr. Iwan Bloch wrote in his book titled *Odoratus Sexualis* in 1933:

> The odor of the secretion from the vaginal passage is related to that of the stinking goose's foot (*Chenopodium vulvaria L.*) as well as to that of cheese, which also belongs to the capryl group. The odor of semen is designated by Haller as odor aphrodisiacus, and is also found in chestnuts and some thorn.

Dr. Bloch also adds,

> The smells of nature are therefore complex chemical combinations which, in gaseous form, possess a high specific gravity. For this reason Zwaardemaker has maintained that natural selection has made use of certain internal properties of atomic structure movements in order to equip the animal organism with sense organs whereby it obtains more information that other forms of life concerning the quality of foodstuffs and the trail of the opposite sex. He has adduced the interesting proof that all animal scents, which influence sexuality, belong to one definite group of chemical relationships, namely, the fatty acids, especially the capryl group. This constitutes the seventh class of capryl odors, or odors hircini, and to which he attributes the disintegrative odors.

I would like to talk about pheromones and their use in perfumes, and by definition, I am referring to a chemical compound produced and secreted by an animal that influences the behavior and development of other members of the same species. First of all, from a marketing standpoint, this is genius. From a personal standpoint, well, I have never had a problem attracting any sexual partner, so for me, it is unnecessary; however, after studying psychology and hypnosis for years, I understand how someone would want this. I have had people tell me that it works, but there is also psychology involved, and I think when someone feels sexy, it goes out into the cosmos, and by law of attraction, bingo, you meet someone who makes you juicy and that you would love nothing more than to bring home with you. So I say, whatever works. The main two pheromones that have been used recently in perfumes are androstenal and androstenone. The third one worthy of mention and that I speak of above is called copulins, and it is a pheromone found in human vaginal secretions. Copulins could be added to the above formula.

Following is an unpublished article written in 2008 by Garry Nelson of Liquid Alchemy Labs of Hawaii, and this may help shed some light on the issue of pheromones:

**Human pheromone—Androstenone (5alpha-androst-16-en-3-one)**

Androstenone is produced by humans, truffles, and pigs. Male pigs or boars to be exact, no doubt this is the reason for the untrue though persistent "pig pheromone in the other guy's cologne" claims that circulate on the internet. This is simply propaganda used by some pheromone companies to scare their customers into only using their products.

Androstenone is found in both men and women. However, it is considered to be a male pheromone as men produce much larger quantities of it. The presence of Androstenone signals high testosterone, sex drive, aggression, and dominance. Androstenone in large amounts can be very intimidating for both sexes and is no doubt the number one alpha pheromone (either alpha male or female) Wearing Androstenone is known for inducing greater respect for the wearer from both sexes.

Around one half of the population cannot smell Androstenone. Of those who can a percentage will tell you that it has a woodsy urine smell or is sweat like , while a smaller percentage will say that it smells like vanilla. Why the vast discrepancy in scent identification? Researchers at Duke University believe that it has to do with genetics. These same researchers say, "In human females, studies show it may cause arousal, sweating and a surge in stress hormones. Women are also much more sensitive to it near ovulation."

An over dose of Androstenone can cause:

- Headaches
- The wearer may smell offensive
- Aggression from other members of the same sex
- An aggressive mood in the wearer, Due to elevated testosterone from the pheromone feedback. Men's testosterone levels rise when they are around men of higher testosterone. Androstenone may be the testosterone indicator.
- Members of the opposite sex may be overly intimidated, Leading to no reaction at all or even a negative reaction. Worse during certain points in the menstrual cycle.

**Alpha-Androstenol (5alpha-androst-16-en-3-ol)**

Alpha-Androstenol is produced by humans and again, pigs. The alpha isomer of Androstenol makes the wearer seem less intimidating and more approachable. People just get comfortable with you.

Alpha-Androstenol is produced by both men and women and does have significant affects on both sexes.

In a test at the University of Warwick, Androstenol exposed men and women both rated women as sexier, warmer, and more attractive than when they were not exposed to Alpha-Androstenol. However, women also rated men wearing Alpha-Androstenol consistently higher on the attractiveness scale while men consistently downgraded other men wearing Androstenol.

Women who are exposed to Alpha-Androstenol tend to initiate social contact with males, and are more likely to be friendly and receptive to males initiating contact with them. Androstenol does cause talkativeness in females, sometimes to the extreme.

Alpha-Androstenol has a musky sweaty aromatic scent. Some say that it smells very much like sandalwood.

An over dose of Androstenol can cause:

*Headaches*
A feeling of being spacey or sleepy, not always a bad thing, some describe it as very pleasant.
An appearance of weakness. In other words, making men appear a lot less alpha, in high doses even fragile.

Following is some food for thought. Several years ago, I was told that when I kissed, the receiver could actually "taste" me being sexually aroused or turned on. I should add this was always at the time of ovulation. That has always fascinated me, and since then, I have been told the same thing several times, so I have to somehow accept that this is fact and is actually happening. When I asked other people whom I felt comfortable with them giving me a straight and honest answer, only the men confirmed this, and it would also happen to them with their wives, girlfriends, etc. Apparently, this did not happen to the women I asked, nor to myself. When I asked the men to describe the taste the best they could, all of them said that it was a salty taste. This is a strange and interesting phenomenon, and if I were researching pheromones, which I am not, I would conduct a more in-depth investigation on this. I have been unable to find any research on this subject. I have added this paragraph intentionally under "Leathers" because I find it to be very "animal," and when I use the word *animal*, by definition, I mean "belonging to the realm of instincts and urges." It is my belief that flavors are perfumes we put into our mouths.

Dr. Iwan Bloch, Sigmund Freud, and Carl Jung would have had a good time with me, or would it have been the other way around?

# CHAPTER 2
# A Day in the Life of a Perfumer

I wrote this article for the magazine, *Perfumer and Flavorist,* which was never published. So here it is.

May 4, 2007

9:00 AM – I am checking my e-mails to see if anything came in since midnight the night before.

9:15 AM – I am at my perfume table thinking of what to do first. Oh yes, I must clear it from the day before and tidy it up a bit. I intentionally do not clear it, just in case something comes to me in my dreams to help me in my formulations. It seems to be never-ending, like the possibilities in an artist's canvas.

As I sip my coffee, I wonder if I added enough of it, Colombian coffee, in my last formula. Only time will tell. I do, however, need to add an ingredient that I have waited for two weeks to get that just came in late yesterday by UPS. Oh well, two weeks is better than two months. I try and stay positive about everything, lest I forget that I am in the middle of the Pacific Ocean on a small island.

I did remember to lock up the formula last night in its little fireproof safe. It is way overcrowded, like a can of sardines, and I have to smash its contents down just to close the safe. I really need to get another vault until all my formulas are scanned into the computers. I had better put that on the list of things to get, and then I had better get back to work on the fun stuff.

9:30 AM – I have added different dosages to seven different bottles, being very careful as I do it. I also put some on my inner wrist to see how fast it dries on me. I usually do at least ten different bottles, but this particular day, I guess I was too lazy to open another box from the storage unit. Besides, I have found that it is best not to get stuck in a routine for it thwarts the flow of creativity for me. Routine is absolutely forbidden in the house of Paragon Perfumes.

Now compounding my own formulas is another topic. I am always shocked to find out that other perfumers do not get their hands and noses in on the action. That to me would be like the artist or sculptor envisioning a work of art in his or her mind and then hiring someone to create it. I just don't get that picture. Besides, I would have just missed out on the olfactory orgasm that I just experienced when I opened the wonderful bottle of juice that I waited for two weeks to get. As I was inhaling this wonderful nectar, I became conscious of myself thinking, while laughing out loud, "Is this legal?"

10:00 AM – On to the next project as I spritz my face with some blue lotus and frankincense. Yummy! OK, just another sip of coffee, Kona coffee, that is.

I clear my table, putting away the glass bottles very carefully. I scan my formula and lock it up safely after meticulously checking my notes, a technique that I learned from researching medicine years ago. I realize my container of lip gloss that I always have on my table beside me is empty, so I open up another one. Ah, another olfactory experience . . . a formula that I made up last year with virgin coconut cream, avocado, Bulgarian rose wax, rose otto from Turkey, vitamin E, etc. I really grew tired of applying stuff on my lips when I do not know what was really in it, which was what inspired me to make this delicious creation.

I then grab bottles of my formulas that I have been nursing for five years. Oh, the patience I must have. As I am reaching for my scent strips, I find myself thinking that this one has grown so complicated. I guess I hang on to it because I do so love a challenge. At the same time, I also think that giving up would be so easy. It reminds me of marriage.

I get the formula out of the vault. OMG! I realize that I have not worked on it since August of last year. That is how busy I have been. Yikes!

I label and date the scent strips and hear a faint drumroll in the background of my perfumed head.

#1 – So-so. As I look at my notes, I recall using some gin in this one. Interesting.
#2 – Nice, just enough wood.
#3 – This is nice, although just about anything with clary sage is.
#4 – Too much violet!
#5 – This floral theme is interesting.
#6 – No comment.
#7 – Citrus theme, nice.
#8 – This is a spicy theme, but it has turned out a little smoky. I added a little Pernod in it. Can you tell that I used to be a bartender? All perfumers should know about alcoholic beverages.
#9 – Oh, I am playing in the woods on this one. I am under a big spruce tree, I am picking mint and chewing it as I take a puff of my pipe with a little tonka-bean-laced tobacco. This one is a nice reminder of growing halfway up in Kentucky. Will my other half ever grow up, I wonder?
#10 – No comment.

10:45 AM – I need a little more inspiration on these, although #9 is close to what I am searching for. Shall I don a mask? I ask myself, as I think of the book *Jitterbug Perfume*, when Tom Robbins talks about the perfumer wearing a whale mask at his perfume table. I guess if I were working on a marine, a water concoction, I would put on my scuba gear. Better yet, I would just go jump in the ocean and swim around with the fish all day.

So I go to my garden instead. First, to the orchid patch, then on to smell the Hong Kong orchid tree, where I pause and think of the orchid perfume that I made last month. How could I make it better? I look down and see the stephanotis clinging on to the bark, wishing it would bloom. I pass the French lavender and grab a handful for my perfume table. I notice a small bloom on the hibiscus and wonder if a little more ambrette might make it wonderful. The cherimoya and magnolia trees look very healthy, but still no blooms. The Paklang trees are loaded with blossoms. I see that the passionflower that I planted at the base of this tree loves it as it is running up the bark and has clung to it like a passionate lover. Excuse me while I indulge myself.

***Passion Flower***

***Hibiscus Bud***

***Hibiscus Flower 'Stolen Kiss'***

***Epidendrum Orchid***

Oh, back to reality. Oh, but wait a minute, this is my reality!

I notice the cereus plants and just can't wait until they bloom next month. I will compare it to my cereus perfume that has been marrying in the bottle. They only bloom one night a year—they are so fleeting, just like life.

I had better get back to the perfumery, I am thinking, as I pass and touch the clove tree, fig trees, the mulberry tree, night-blooming jasmine, the coffee tree, crotons, virburnum, the kaffir lime trees, ferns, etc. I will have to go into the woods on a hike up into the mountain and smell the redwoods, the koa, and the eucalyptus trees. There are so many scents to choose from.

***Coffee Blossoms***

***Snowball Virburnum***

***Crotons in bloom***

***Brown Turkey Figs***

All perfumers should have a deep understanding of plants and their symbiotic relationships.

11:00 AM – Time for brunch.

12:00 PM – I was thinking as I was immersed in my culinary delights that perhaps my chef background might inspire me, but no, not today. Every perfumer should know flavors.

Oh, I wish I could just make a phone call to Gattefosse. He would probably say, "Just add some more clary sage, my petite." Thinking of the phone, I should check my messages to see if there are any dire emergencies, because it would take one to drag me away from my table.

I remember to sniff my inner wrist for the dry out. I always write the time on myself too when I do this, so by the end of the day, I look like a tattoo artist.

There is one important e-mail from a company that I deal with, advertising their raw material collection. Oh, I like the sound of that. I'm in, sign me up, and send them on over. The other important one is from a retail store wanting a rush order for Mother's Day. I guess I will have to ship it overnight. Good thing it is on a neighbor island. Oh yes, I just remembered that I have to ship a perfume sample to a company to be tested for eight weeks. I guess that I will have to run to the post office on my next break.

12:15 PM – I had better warm up my resins because I have a feeling that I'm going to be using them later today. OK, now I need to check the blotters on my other table. They have been there for one month, and I need to test the dry out.

So far, I'm happy with some greens that I've been working on. I am surprised that I like them, because they're the least of my favorites. But on first smelling them a month ago, I was transported to the year 1992, in Napa Valley, on a wine-tasting trip with some friends while in medical school. Later, I will find those notes and reminisce for some inspiration on how to make my Chardonnay-themed one even better. Next, the ambers—yummy, you must love ambers!

12:30 PM – Now what do I do next? I was just looking at my to-do list, and eureka, there it is, the ingredient that I want to add to my five-year-old formula. I had even made a little note to myself. I guess this is where my herbalist background comes in. That is what I get for having too many irons in the fire. Damn you, Gemini moon!

OK, back to the drawing board, so I grabbed the material that I had macerated back on September 14, 2006, as it should be perfect by now. But how much should I add? This seems to always be the question. And which bottle shall I add it to? I might have to channel Jacques Guerlain, as I think he might have used this ingredient in several of his formulas. Such a master perfumer, he was.

I am feeling a bit ornery, so I will overdose bottles number 6 and 10. That is exactly the feeling that I was searching for today. I have found that it is important for perfumers to be in touch with their feelings and to get out of their heads. Perhaps, there should be a sign at my door that reads, "Upon entering Paragon Perfumes, all perfumers please leave your head at the door." I imagine Jacques was feeling very ornery when he made his famous Shalimar.

I like this ingredient so much that I am tempted to add it to my green family. The more I inhale it, the more I like it. OK, I am going to put it in. Like I said, I am feeling ornery. Now I must find the bottles and the formula from the vault.

I added just a minute amount this time. I guess I'm back into my head now. This was a complicated formula for me, and I challenged myself by saying, "No woods, no fougere, no fruit, no musk, no powdery dry out, no marines, no spice, not too floral, and good luck to myself." I got very lucky, however, because bottle number 2 came out really nice on the first trial run. That is just sheer luck, I believe, or maybe just amazing perfume karma, if there is such a thing.

1:20 PM – Back to my to-do list. Now that I have my ingredient that I've waited two weeks for, I can add it to all of the lonely formulas that I'd made previously. They've been waiting for it. This is going to take a while; I will start with the plumerias.

***Honey Girl***

I remember to take a whiff of my inner wrist. Yummy!

2:30 PM – I have finished the plumerias. I already had written in my notes how much to add; it was just time consuming. Next, I'll add it to my lime blossom and lotus flower. First, a little nose nudge from Honey Girl, my dog and Paragon's mascot. She lets me know it is time to take a break and . . ."What's that, Honey Girl, it is teatime too? Oh, you're such a smart dog . . . you must be Lassie's offspring." After all, she is part collie. I suppose it's appropriate to have a dog instead of an obnoxious alarm clock. Google Inc. also allows your dog to come to work with you as long it's a cool dog. Yet another idea for a new sign at my door that reads, "Only cool dogs allowed."

2:45 PM – So I get up and spritz my face with a little immortelle face spray that I made. Yummy, will this really make me immortal? How about younger-looking? Only time will tell. Now for a little tea and water with some lemon balm hydrosol, a breath of fresh air, and then back to the florals.

As I stand outside the perfumery completely mesmerized by the beauty before me, seeing the vast ocean and majestic mountain, I am reminded of why I chose to live in this beautiful paradise, Maui.

3:00 PM – Back to work.

3:30 PM – I am finished with the lime blossom and lotus flowers. Now I must spend at least an hour on scanning formulas into the computers. These are for safekeeping and for my book that I'm working on, where I will reveal my perfume formulas. This is a very boring task, so I will play some iTunes to get me through it. I think some belly dance music would be good, or how about some ballroom?

4:30 PM – I have finished the scanning, and now I need to check the e-mails again. I have to answer a company wanting our UPS account number to ship some samples. Where is that account number? Is it too late to run to the post office?

5:00 PM – Now to start typing this since I've handwritten it as I worked. Tomorrow, I suppose I will work on some lavender, and then off to the ball I go, where I am sure the waltz will inspire me. On Sunday, I'll go deep into the woods. I had better not forget my notebook.

6:00 PM – Time to take a break and go for a one-hour walk.

7:00 PM – Dinnertime.

8:00 PM – Back to typing.

9:30 PM – Finished typing finally. I can smell the resins on the warmers. Is it too late to make some more perfume?

## CHAPTER 3

# The Intoxicating Tree

I will never forget the first time that I smelled this intoxicating flower blossom. I had followed this amazing scent trail with my nose for quite some time while strolling around in a nursery, savoring every breath, until I literally walked right into a branch sticking out around a corner where someone did not prune it. Laced along the branch was the most enchanting flower blossom. I stood there, completely captivated and in a state of pure olfactory ecstasy. All time had stopped, including my heartbeat, my breath, my sense of touch, my vision, and all of my thoughts, except for my sense of smell. In a split second, I was taking a sensual journey and became one with this beautiful flower. My mind was wrapped around this simple little blossom and all of its essence. It was all of the pure positive aspects of love, being loved, being in love, and being the beloved. How could a little flower have such power over me? Or was it that I had all the power over it? I was wholly (holy) hypnotized into the "real world." Everything else is an illusion. This I believe with all of my being.

After a long time, what seemed like eternity, a young man came to help me, and I asked him about this tree. He said that they had just come in that morning, all in full bloom, and there were ten of them. He also said that he could not stand too close to them downwind because their scent was so intense that it made him want to vomit, to which I responded, "Perfect, I will take all ten!" That was all that I needed to hear to realize that this tree was worth investigating.

This beautiful flower of the *Michelia champaca alba*, grows on a small tree belonging to the genus *Magnoliaceae,* also called the Chinese violet, pakalang, paklang, or paklan, and is found in Hawaii. Since I believe there are no indigenous plants (expect for possibly the silversword) or trees from Hawaii, there is some curiosity about where paklang came from. My best educated guess would be that it was originally grafted in Thailand or China from the *Michelia champaca* and the *Michelia longifolia.* Out of all of the nationalities I questioned that live in Hawaii, all of them denied knowing anything about the tree. My first guess was that it came from the Philippines, but the Filipinos say they have never heard of it. I do know that the *Michelia champaca* does exist in the Philippines, Madagascar, and on Réunion island. I invite my readers to write to me regarding any information on this tree since my research on the internet has not shown to be one of much great detail.

***Paklang Blossom***

Poucher makes mention of the paklang and has it having a violet type odor and has it listed in his second volume under the name *Pergularia odoratissima.* I don't believe that he could have ever smelled the paklang flower blossom because it smells nothing like a violet, even though its nickname is the Chinese violet supposedly. The paklang's scent is very hard to describe; it reminds me of one of my favorite orchids, the *Encyclia cochlea*, which I will speak about in another chapter. It is sweet and sugary, but jasmine-like with a fruity undertone. It is also not white, as the name *alba* would imply, but rather an eggshell off-white color, to be more specific. It is a starburst, and not shaped like a tulip,. It smells nothing like the *Magnolia grandiflora*, the lovely tree indigenous to the southeastern states of the USA, also belonging to the same family. I believe Poucher may have been talking about the fragrant Telosma or Tonkin creeper because the scientific name is *Pergularia odoratissima*, and it is in the Asclepiadaceae family.

The oil from the *M. champaca* contains phenyl ethyl alcohol, benzoic acid, benzyl alcohol, benzyaldehyde, *p*-cresyl methyl ether, isoeugenol, linalool, *cis*-linalool, benzyl acetate, phenyl ethyl acetate, phenylacetonitrile, eugenol, cinnamic alcohol, ionones, benzyl benzoate, phenyl ethyl benzoate, methyl linolenate, cineol, and other minor and trace components. The main components of the concrete are indole, methyl anthranilate, eugenol, methyl linoleate, methyl palmitate, phenylacetaldoxime, theaspirane A, *cis*-linalool oxide, phenyl ethyl alcohol, methyl benzoate, methyl phenyl acetate, and methyl benzoate. The *M. longifolia* contains methyleugenol, linalool, methylacetic acid, a phenol having the odor of thymol, and probably a myriad of other amazing constituents.

Unfortunately, I do not have the headspace information on this flower; however, here is what I do have. Since I was unable to obtain the flower oil at the time of the testing, I made my own by extracting it from greases, the painstaking, old-fashioned way. I used blotters at a 1:1 ratio on all tests, and all raw materials are authentic unless indicated otherwise. Following is what I discovered.

The following raw materials blend well with this flower:

Labdanum
Geranium
Lemon
Green lemon
Chamomile
Neroli
Cognac
Benzoin
Rosewood
Frankincense
Orchid, synthetic
Pikake
Magnolia
Narcissus, synthetic
Mimosa absolute
Elderflower
Elderberry
Freesia, synthetic
Cocoa
Lotus (blue, pink, and white)
Jasmine
Mango
Myrrh
Smoky myrrh
Vetiver
Cypress
Opium, synthetic
Nag champa, synthetic
Elemi
Cedarwood
Nasturtium
Passion fruit (a.k.a. lilikoi), synthetic
Tuberose

Amyris
Balsam fir needle
Ho wood
Palma Rosa
Angelica root
Opopanax
Dragon's blood
Lavandin
Spike lavender
White truffle
Black truffle
Nepalese spikenard
Muguet, synthetic
Scotch whiskey
Immortelle
Melissa
Oud (a.k.a. agar wood)
Ambrette seed
Beeswax absolute
Bergamot
Mint bergamot
Black currant seed
Canadian cedar leaf
European cedar leaf
Canadian western red cedar leaf
Erigeron (a.k.a. fleabane)
Moroccan mugwort (a.k.a. armoise)
Oakmoss
Orris root
Tagettes
Bulgarian wormwood (*Artemisia annua*)
Ylang extra

Cade
Ledum
Rhododendron
Turkish rose
Moroccan rose
Sea buckthorn berry
Boronia
Tangerine
Satsuma tangerine
Juniper berry
Costus
Davana
Honeysuckle, synthetic
Pink crinum lily (extracted by greases)
Sandalwood
Tolu balsam
White sage
Sweet yellow birch
Copaiba balsam
Horehound
Bitter Orange
Tarragon
Mace
Farehitra
Yellow mandarin
Clementine
Savory
Coriander
Lantana
Borage seed
Reishi
Amber
Walnut liquor

***Freesia***

***Iris***

***Navel Orange Blossom***

***Lantana***

The following raw materials do not blend well with this flower:

Osha root
Oregano
Lovage root
Guaiac wood
Vanilla
Cilantro
Valerian
Sweet flag
Pine moss
Pua keni keni
(a.k.a. *Fragrea berteriana*)
Peach leaf absolute
Mugwort (Douglas from the USA)

Manuka
Musk rose
Choya loban
Lemongrass
Fennel
Wild birch
Marjoram
Rosemary
Thyme
Bay
White spruce
Spearmint
Cinnamon leaf

Cardamom
Anise
Allspice berry
Patchouli
Peru balsam
Tea rose, synthetic
Clove
Carnation
Osmanthus
Basil
Lavender
Lime
Cream sherry

The following materials are just OK or so-so with this flower:

Lemon verbena
Gardenia, synthetic
Plumeria, synthetic
Peony, synthetic
Crepe myrtle
Petit grain
Hemp
St. John's wort
Grapefruit
Clary

White camphor
Poke
Niaouli
Chambord
Macadamia nut liquor
Frangelico
Nutmeg
Kahlua
Sweet birch
Cajeput

Catnip
Violet leaf
Yarrow
Yuzu
Calophylum
Roman chamomile
Hay
Red thyme
Black spruce

***Orange Blossom opening up***

***Hedychium gardnerianum***

I do not, at least at the time of writing this book, have a good source for this oil, so I make it myself like I mentioned earlier. Needless to say, it is very time consuming, but well worth it.

***Paklang blossom with buds***

# CHAPTER 4

# The Wonderful World of Orchids

***Cymbidium hybrid***

There is an incredible abundance of orchids on this planet, and botanists are still discovering more. The end of 1920, there was recorded over six thousand species; and by 1990, there were over twenty-five thousand. That's approximately twenty thousand in less than a century. There are actually species of orchids that grow on an entire cliff side that are protected by deep gorges that you can only view from afar and never be able to smell up close. They are untouchable, therefore unable to record. Orchids comprise 10% of all of the plants on the planet. The orchid flower family is my favorite, probably because they are so mysterious and elusive, and I have been fascinated by them all of my life. Hawaii is now loaded with orchids that enthusiasts have brought to the islands, and I am fortunate to live in such a place to be able to study them.

In the past, there has been some controversy over orchid enthusiasts that have smuggled orchids into other countries (including the United States) and have gone to jail and served time in order to grow and study them. I would like to thank those people who have smuggled these orchids into these other countries because in hindsight, the same people that jailed these individuals also watched large plots of land, forests, and jungles full of rare species of orchids burn in flames (at their hands) and perish into ashes just to be able to clear the land for roads and the raising of cattle.

And they have done it *knowing* that it would wipe out an entire species. When interviewed about their actions, their response was, "So what, they are our orchids, and we can do whatever we want with them." This kind of response is usually indicative of an ignorant third-world country where there is lack of education. It is most unfortunate. Thank goodness those smugglers had foresight; otherwise, some of the orchids would have been completely wiped off of the planet, and a whole lot were. If you are one of those persons that did jail time, thank you. You have done a great deed for humanity.

One of the most educational pieces of work done on the scent of orchids is by a man named Roman Kaiser, and he produced a book called *The Scent of Orchids*, published in 1993. This book, as well as Roman, will go down in history, no doubt. The book gives the detailed chemical analysis of orchids, has beautiful photographs, and talks about the botanical and pollination aspects. It is a masterpiece.

My absolute favorite orchid perfume, besides my own, is Nuit d'Orchidee by Yves Rocher. Another one was called Orchidee, but I am not as crazy about that one. They both came in beautifully colored round bottles with a nice sparkle top that reminds me of lovely Arabian architecture. The Orchidee is clean and fresh, but the Nuit d'Orchidee is exquisite, and I think that it is a perfectly balanced perfume. I have both in my collection, but I do not believe these are being manufactured anymore. Neither one smells anything like an orchid that I have ever smelled, but I like them just the same.

There is a black orchid, but it is scentless. It is the orchid genus *Dracula* and consists of 118 species. The strange name *Dracula* literally means "little dragon," referring to the strange aspect of the two long spurs of the sepals. They were once included in the genus *Masdevallia*, but became a separate genus in 1978. The genus has some of the more bizarre and well-known species of the subtribe Pleurothallidinae. When I was in Panama, I was fortunate to be able to see these growing in the wild. There is something rather spooky about them.

To my knowledge, there is not one recorded orchid that is either toxic internally or externally. Nor is there any known orchid that causes hallucinations, so I say, if you can extract that awesome juice, then go for it. There is an orchid species that is made into a food; it is known as salep, and medicinally, it is a demulcent. In the Orient, it is used as an aphrodisiac. When I was traveling in Istanbul, Turkey, I remember having the most delicious pudding made with this orchid. I also had it at a Middle Eastern restaurant in Vienna, Austria, that specialized in presenting the most wonderful belly dancers from Egypt. Classical Egyptian dance is another one of my passions besides perfume.

***Ladyslipper***

Here are some of my favorite orchids:

*Angraecum eburneum* – I absolutely love this orchid, and the aroma it emits to attract its beloved night moth, with its sixteen-inch tongue. I have met other people in the orchid society, however, that said that the smell of this orchid made them want to vomit. Oh well, to each their own. I wasn't at all surprised to find that the most dominant constituent in the headspace analysis is benzyl acetate, one of my favorite raw materials to play with. Call me crazy, but I almost love this chemical as much as I do authentic jasmine, pikake jasmine, and night-blooming jasmine.

*Angraecum eberneum superbum* – I was actually surprised by the analysis of this orchid showing a dominant content of (Z)-3-hexenol. My guess would have been dominant benzyl acetate; however, it only shows a trace in this analysis. I do appreciate the smell of fresh-cut grass.

*Angraecum eichlerianum* – This orchid has the beautiful aroma of jasmine with the dominant constituent being benzyl acetate. The leaves of the *angraecum* species are used for making tea in Madagascar.

*Zygopetalum mackaii* – This wonderful orchid has an aroma of a bouquet of hyacinth mixed with wisteria and lilac. Unfortunately, I was unable to find the headspace information on this one and wouldn't dare compare it to its related cousin, the *Zygopetalum crinitum*.

***Zygopetalum 'Bluebird'***

*Cattleya labiata* – Reminds me a little of a pink crinum lily mixed with hyacinth, wisteria, and lilac. From what I can gather from research, my guess is that they were originally found in Rio de Janeiro in the early 1800s and ended up in the Kew Botanic Gardens. Here is a formula called the Kew Garden Bouquet that I would like to think was made to mimic this orchid. It comes from the book from 1877 called *A Treatise on the Manufacture of Perfumes and Kindred Toilet Articles* by John H. Snively, PhD, a professor of analytical chemistry in the Tennessee College of Pharmacy, also author of *Tables for Qualitative Chemical Analysis*. This man did his homework.

**Kew Garden Bouquet**

Orange flower spirit ........ 2 oz.
Rose geranium extract ........ 1 oz.
Cassie essence ........ 1 oz.
Tuberose essence ........ 1 oz.
Jasmine essence ........ 1 oz.
Musk tincture ........ 6 fluidrachms
Ambergris ........ 3 fluidrachms

Keep in mind that all of these materials were authentic, so when it was first made, it must have been fabulous.

Perfumers' still use salicylates when making orchid perfumes; however, this is an old school of thought, and it is unnecessary to use this old way of thinking when designing a modern orchid perfume. Nowadays, you can use any of the thousands of available raw materials that are on the market. Also there is not one orchid perfume on the market that I know of that smells anything like an orchid, so it really doesn't matter anyway. They are all pretty much *fantasy* perfumes and were originally made as a marketing ploy. I have made orchid perfumes that smell exactly like orchids but they are not for sale as of yet.

Here is one of my renditions of an orchid perfume by P. Jellinek from 1949.

**Orchid by Bobbie Kelley**

|  |  |
|---|---|
| Amyl salicylate | 110 |
| Linalool | 30 |
| HDXC | 50 |
| Ylang | 80 |
| Heliotropin | 60 |
| Coumarin | 30 |
| Ambrettolide | 20 |
| Oakmoss | 10 |
| Jasmine absolute | 10 |
| Aurantiol | 10 |
| Hyacinth body BHT | 20 |
| PEA | 50 |
| Vanilla | 10 |
| Benzyl acetate | 40 |
| Neroli | 30 |
| Bergamot | 80 |
| Tuberose | 10 |
| Patchouli | 10 |
| Vetiver | 10 |
| Tolu | 40 |
| Tonka bean absolute | 10 |
| Anisic aldehyde | 20 |
| Green lemon, terpeneless | 20 |
| Sweet orange, terpeneless | 20 |
| Lilac | 120 |
| Amyl cinnamic aldehyde | 40 |
| Ant methyl | 10 |

***Vanilla Orchid***

This is not the orchid perfume formula that I have on the market.

*Cattleya warscewiczii* – A dazzling orchid from Colombia; what else can I say?

*Cattleya jenmanii* – An incredible orchid that was discovered in Venezuela in 1964.

*Cattleya schilleriana* – A divinely scented orchid with a strong sweet odor with geraniol being the dominant factor.

*Lycaste skinneri alba* – This orchid blows my mind as well as my pocketbook; I have bought way too many for $100 a pop, only to watch it die. But they are so lovely and, I believe, one of the most heavenly scented orchids in the world. They are from South America and are amazingly rare and beautiful. And yes, after it bloomed, it died, but its scent is still implanted in my olfactory storehouse of scents. Its aroma is dominated by methyl (E)-cinnamate.

*Brassavola nodosa* – A delicate-scented orchid that resembles a rosy bouquet of flowers with the dominant constituents being (E)-ocimene, geraniol and (E)-nerolidol. This orchid's nickname is Lady of the Night.

*Brassavola tuberculata* – This lovely orchid reminds me of a mix of pikake jasmine and gardenia with (E)-ocimene, methyl salicylate, and phenyl ethyl acetate being dominant.

*Brassavola digbyana* – Lemon-like.

*Encyclia pentotis* – Also known as *Encyclia baculus*, this orchid is dominated by (E)-ocimene. The scent of this orchid is very hard to describe; the first thing that comes to my mind is that it smells like candy from outer space. If this orchid were candy, I would eat the whole plant.

*Encyclia cochlea* – See above description, plus, it is loaded with benzyl acetate and honeysuckle. Now that would be some awesome candy!

*Encyclia fragrans* – Found in Central and South America, it smells similar to a tropical fruit and was also at the Kew Botanic Gardens.

*Stanhopea wardii* – I was lucky to smell this wonderful orchid during its short seventy-two-hour lifetime, and it was fabulous. I have tried to grow this orchid also, and I have failed. I have also smelled the *Stanhopea ecornuta*, which looks like a butterfly, and the *Stanhopea jenischiana*, which looks like a hummingbird. They are truly magnificent.

*Stanhopea tigrina* – I will never forget the first time I saw and smelled this awesome and overpowering orchid. It has spots and stripes like a tiger and smells like a bucketful of phenyl ethyl acetate, another one of my favorite chemicals to sniff. Apparently, a particular bee walks around on the bottom of the orchid, gets intoxicated, and then falls inside of what looks like a little cage. Then he has to find his way out at the bottom, thereby stumbling upon the pollen, scraping it along his back or belly, and then he repeats the process all over again on the next one. Oh, the life of the bee. The phenyl ethyl acetate was recorded at 92% on this orchid, which did not at all surprise me, followed by only 1.6 of methyl salicylate and 1.4 of benzyl acetate. I must admit I have not had the heart to tear this orchid apart, probably because I don't have to because its scent is almost overwhelming. I guess I want it to be like that, and I want it somehow to be mysterious and leave me intrigued, where I am constantly guessing what else is in it, even though I can see the headspace technological statistics right in front of me on paper. I want it to somehow laugh at me and be saying something like, "Ha, you will never be able to figure me out." If you turn this orchid over and look at it from the backside, it looks just like a mask; it is a beautiful sight to behold, and very unique.

***Stanhopea tigrina***

***Stanhopea tigrina***

***Stanhopea tigrina***

***Stanhopea tigrina***

*Bulbophylum medusa* – A stunning orchid that smells of sperm.

*Bulbophylum phalenopsis* – An eye-catching orchid that reeks of rotting meat accompanied with maggots.

*Bulbophylum macranthum* – Smells like cloves.

*Bullbophylum vitiense* – Smells like coconut.

*Maxillaria tenufolia* – This orchid is a little challenging to grow and also smells like coconut. It is a lovely dark orange to red color.

*Dendrobium moschatum* – Smells musk-like.

*Odontoglossum citrosum* – Lemon-like.

*Oncidium sharry baby* – This orchid smells good enough to eat with its chocolate scent laced with a touch of vanillin.

***Oncidium sharry baby***

There is another note in orchids, and most other flowers for that matter, that have a very distinct smell of indole, civet, skatole, etc., especially when they are wilting or dying. This smell is sometimes dominant in my olfactory senses, but is not recorded in the headspace technology. It is my theory, then, that indole is partly responsible for the natural decay of some flowers, and that is why I believe there should always be something a little *stinky*, for lack of a better word, in a floral perfume to make it more natural and seem "real."

I do want to mention the Hong Kong orchid tree, or Singapore orchid, that is not an orchid at all, but some people think that it is. The flower is a beautiful magenta-colored pink, and the smell is very delicate and does remind me of an orchid. See the photo and the chapter on dissecting flowers.

***Pink Cymbidium Hybrid***

# CHAPTER 5
# Dissecting Flowers / Single-Note Floral Perfumes

***Camellia***

I love everything about flowers. I think that they are beautiful throughout their entire life. There is something about how they change in size, color, and scent, and how they are ever flowing even as they wilt. Their life is truly magical, and even though it is short, it is full of surprises. So it pangs me so to tear one apart, but this is how I learn to distinguish one from another. I wish it were easier than this, and I am envious of the bees and moths that just home in by scent and sight. And as much as I love headspace technology and think that it is incredibly helpful, I really have to trust my nose more than anything else, and I believe that goes for all the rest of the perfumers of the world. I have my own little unique way of sniffing flowers, and I can't speak for other perfumers, but this one works for me. I call it dissecting flowers. I actually use a scalpel and forceps (same thing as tweezers) and go to work on the helpless blossom, and then I build a perfume based on my senses. Make sure the scalpel does not have any other scent on it, and if you are unable to get scalpels, you can use a box cutter, knife, etc.. Just make sure it is sharp.

I could probably write volumes on flowers, and actually, volumes have been written about them; so instead, I have written the main highlights about the flower that I am analyzing or what I feel is important when creating a formula. I will begin with one of my favorite blossoms, the lime blossom. And by the way, in my note taking, I write everything down, even if it sounds crazy, strange, different, etc. It is important to write all thoughts, as this is a part of the creativity process.

## The Lime Blossom

When I first smell the lime blossom while it is on the branch, there is a blast of fresh citrus, followed by dominant aurantiol and indolic notes. I am a little surprised by this but try not to judge its character. It is what it is. There is also a lily of the valley scent that is delicate and perfectly balanced. Immediately after plucking the flower from the branch, there is a very strong woody and nutty odor. In the base of the blossom, there is a very strong citrus odor that is different from the initial scent while still on the branch. This particular scent is stronger, with a green note. After cutting the delicate petal with the scalpel, I discover that deep in the heart of it is a jasmine-like note, smelling fatty and lactonic. Even though there seems to be no dominant citrus note at the top of the blossom, when I squeeze on the petals and stamen with forceps, there is a wonderful lemon scent that reminds me of the orchid, the *Odontoglossum citrosum*, that possesses a delicate lemon-like odor that I would describe as fabulous. Also, the orchid *Brassavola digbyana* comes to my mind (and nose), in that it too has this delicious, almost-indescribable aroma.

It is interesting to note that not all of the blooms smell alike, neither the concentration nor consistency. It seems the blossoms are somehow tricky and appear to be laughing at me trying to figure out their secret scent. I attribute this to the fairies dancing around in my face. Remember, I said, no matter how crazy it sounds. To be more realistic, it is probably more having to do more with the time of day and the rates of evaporation into the ethers of the universe. The blossoms were picked at 10:00 AM Hawaiian time and, in hindsight, probably should have been dissected a lot earlier.

So the raw materials that immediately come to my mind are the following:

Cinnamic alcohol
Aurantiol
HDXC
Indole
Lime
Lemon
Green lemon
Orange
Grapefruit
Toscanol
Nerolidol
Vetiveryl acetate
Vetiver
Benzyl acetate
Hedione
Amyl cinnamic aldehyde
Ant methyl
Pittosporum
Citral
Tonka and/or coumarin
Sandalwood
Civet
Farnesol
PEACE

After one hour and forty minutes, the blossoms start smelling very sweet and sugary. One ingredient that comes to mind that is similar in a chemical kind of way is phenyl ethyl acetate another is cinnamic alcohol. Lastly, two more ingredients come to mind, and they are allyl amyl gluconate (for its honey-like character and its molecular weight of 186.2) and honey absolute (for its cloying sweetness). With that being said, now ants surround the blooms; they are attracted to sugar.

Here is a lovely lime blossom by Jellinek.

**Lime Blossom No. 1094**

| | |
|---|---|
| HDXC | 400 |
| Terpineol | 150 |
| PEA | 180 |
| Ylang | 20 |
| Limes, terpeneless | 10 |
| MNK | 50 |
| Iso-jasmone | 30 |
| Citronellyl acetate | 70 |
| Methyl octine carbonate | 5 |
| Isobutyl phenylacetate | 10 |
| Aldehyde C12 | 5 |
| Jasmine abs. | 20 |
| Broom abs. | 10 |
| Heliotropin | 30 |
| Musk ambrette, authentic | 10 |

This formula has depth and is rich and intense. Of course, there is both authentic jasmine and broom absolute, which would really contribute to the depth, and musk ambrette, which makes anything smell delicious. Linden blossom absolute blends well with ylang, rose, neroli, jasmine, verbena, grapefruit, amber, violet, ginger, benzoin, palmarosa, and lemongrass. There is a synthetic linden blossom base made by Ferminich International and sold by Vigon International called Linden Blossom Givco 151 PCMF, and it is a good try; still, there is nothing like the real thing.

## The Cereus Flower

I will never forget the first time I smelled a cereus flower. I was finally able to find some growing along the road and waited until midnight (they are nocturnal bloomers) to go and pick some to dissect, and of course, as soon as I parked there, the police pulled up and wanted to know what I was doing out on the road in the middle of the night. It wasn't long before he let me pick as many as I wanted, and he also told me that they are such a nuisance here on Maui and are so overgrown on the side of the road that the county has a hard time every spring cutting them down. They are in the cactus family and are very prickly, so no one wants to deal with them. Following are my notes from that full-moon night, and I have no idea what exact species these were.

***Night blooming Cereus***

*Surprisingly, the petals smell like milk chocolate, specifically, not dark chocolate, with an underlying somewhat fatty, lactonic, and waxy aroma. It is unlike anything that I have ever smelled before, which makes it so intriguing to me. There seems to be no trickiness about them, all smelling alike. The piston smells floral, slightly woody and green. It is making me sneeze, which is very unusual for me since I have no known allergies to any plants other than the white mulberry tree, for which I was clinically tested. It also smells slightly watery and a little metallic like aloe vera. If I stick my face deep inside the middle of the flower, I smell a nutty odor that I have never smelled before in a flower; it is very pleasant and makes me want to eat it as I continue sneezing. The only nut that I can think of is the macadamia nut because it smells so fatty. I can also detect a very minuscule orris note. These odors remained to be dominant all through the night, until the flower finally wilted, then perished. I did not ever smell any spices, fruit, musk, animal, cheese, flesh, honey, powder, fishy, citrus, vanilla, rose, jasmine, or any other flower odors, balsams, stinky or rotten, roots, sea water, herbs, or vegetable odors.*

I have both a Night Blooming Cereus perfume solid on the market and a Night Blooming Cereus parfum.

## The Paklang

When I smell the blossom on the tree, the first odor that I detect is a wonderfully sweet orange blossom, with a deep, rich blood orange hidden below, almost as if an orange blossom is growing out of a blood orange that has not yet reached maturation. The only other flowers that come to my mind are the gardenia, the orchid called *Encyclia cochlea*, and the mimosa. After plucking the blossom, the above odors dominate; however, it appears now that a faint lilac aroma starts to ascend from the middle. When I smell the stem and squeeze it with forceps, there is a very dominant green, which is very overpowering with what seems to be linalool from the rosewood tree. There is a deep and profound sweetness to the blossoms as a whole, and as I search for words, my olfactory pathways are penetrated by this delicious, almost-sickening, strange, and unfamiliar sugary sweet odor. A flavor that comes to my mind is icing on a cake, perhaps a buttercream type, and it is always so tempting for me to not take a big bite as I am sniffing. As the delicate blossom starts to fade, there is a dominant cheesy flavor mixed with indole and a ground black walnut paste, which is quite nauseating.

***Paklang***

At the present writing of this book, I have been working on a paklang formula. See chapter three which is dedicated to the paklang.

## The Hong Kong Orchid Tree

***Hong Kong orchid blossom***

Some people have mistaken this flower blossom to be an orchid, which it is not. On the tree, this blossom smells of an unknown flower mixed with PEA and linalool; and as soon as I pluck it from the tree, the odor practically disappears. After cutting the beautiful magenta-colored petal, linalool dominates, followed by salicylates, then violet. When I crush the stamen, linalool once again dominates, with an underlying green note, followed by a sweet balsamic aroma, and, lastly, a very slight coumarin odor. The hibiscus flower comes to mind. The ingredients that flood my thoughts are, of course, high quantities of linalool first, then PEA, salicylates, ionones, styrax, benzoin, and the tonka bean.

I don't know of any Hong Kong orchid perfumes on the market in the United States, nor am I working on one at the present time; but I do have a tree in my backyard, and I very much enjoy sitting under it as the butterflies surround this tree like bees swarm around honey. It is captivating.

At this point in this chapter, I would like to mention a special collection of ingredients that was presented to me by Vigon International—called the Collection—that I have enjoyed playing with and that a perfumer could incorporate in any of these floral concoctions that might add a certain fanciness to the blend, and I have had a lot of fun testing these synthetics.

My notes follow verbatim:

1. Ebanol (Jasmine) – OK, nothing great, smells cheap.
2. Ebanol (Licorice) – Not bad for licorice, might be good in a hyacinth or chypre.
3. Florhydral (Muguet) – Smells like cyclamen aldehyde and HDXC mixed with benzyl acetate. This is supposed to be a muguet?
4. Florhydral (Citrus) – This is supposed to be a citrus? It might be good in a soap powder or laundry detergent.
5. Javanol (Gardenia) – At last! This one is very nice and there is a deep richness to it. Hats off to you, chemists!
6. Javanol (Sandalwood) – This smells like a cheap brush cedar mixed with a cheap terpineol and pine-sol. This is supposed to be sandalwood? Could this possibly be an old and expired synthetic?
7. Labienoxime-10 (Grapefruit) – Can you guys think of easier words to say? This is supposed to be grapefruit, but I smell Tangerine and Kumquat and it is very intense. This has great potential.
8. Labienoxime-10 (Sweet peach) – Come on, you guys, can you be a little more creative on thinking of new words instead of repeating all of these? When I first smelled this, I wanted to vomit! It is so sweet; therefore, I think it would be great in a perfume for teens. Again, hats off to the chemists!
9. Methyl diantilis (Carnation) – It smelled sort of weak on the blotter; however, I accidentally spilled some on my fingers, and I was unable to wash it off even after several hand washings and a shower. I consider this synthetic to be very powerful.
10. Methyl diantilis (Vanilla) – This is that yucky, cheap cotton-candied synthetic vanilla that I can't stand. It would probably be great with that sweet peach (number 8) for a fragrance for teenagers.
11. Methyl laitone (Chocolate) – This is like a sweet cup of chocolate milk that has been sitting on a table all night long next to a chocolate chip cookie. It's all right.
12. Methyl laitone (Agrestic lavender) – This is supposed to be lavender, but smells like soap with camphor and underlying balsamic and green notes. Smells clean.
13. Okoumal (Musk) – This is bizarre. This is supposedly musk?
14. Okoumal (Tobacco) – This is OK for a woody amber tobacco.
15. Peonile (Mimosa) – It smells very nice, light and floral, although I would never guess this is a mimosa.
16. Peonile (Red fruits) – This is bizarre and smells like laundry detergent.
17. Pharaone (Piña colada) – It is nice for a piña colada; I like it. It makes me want to go sit on the beach and sip on a nice cold piña colada.
18. Pharaone (Red apple) – Smells more like an Asian pear to me. It's very nice, actually; I like it.
19. Safraleine (Leather) – Animal and mossy. This begs to be put in Russian leather.
20. Safraleine (Spice) – This one smells like carnation, cloves, and pimentos and reminds me of classical perfume Old Spice.
21. Toscanol (Green leaf) – This one smells like a citrus leaf and could be used in any of the citrus blossom formulas.
22. Toscanol (Ylang) – I would not have ever guessed this was ylang.
23. Ultrazur (Cologne) – I smell elemi and linalool.
24. Ultrazur (Pear) – OK for a pear.

## Magnolia

***Saucer Magnolia***

Back to dissecting flowers. Allow me to expound on one of my absolute favorite flowers, the *Magnolia grandiflora* Just saying the word to myself conjures up visions of me being in the Deep South, sitting under a huge magnolia tree, home to long hanging tendrils of toxic moss and beautifully colored bromeliads that glisten as the sun is setting. Slowly swinging naked in a rope swing tied tightly to her branches, whilst I smell the bayou in the distance, I can hear soft *coon ass* music playing as I sip my mint julep. Oh, where was I? Back to dissecting the flower . . .

When I smell the magnolia flower while on the branch, I smell the following odors: honey, lemon, rose, muguet, cinnamon, frangipani (the stamen), greens, sweet wood, gardenia, jasmine, farnesol, lactones, cyclamen, narcissi, freesia, lily, aurantiol, nerol. The magnolia flower, after being cut from the branch at various stages as it is dying, has a distinct watery smell, and there is also a stinky smell, like rotting flesh. And just when the one blossom was almost dead and looked very brown and dried up, there was a wonderful lemon verbena smell that lasted all throughout the night. There was another younger blossom that was starting to get a cheesy smell as I sat up all night sniffing.

Following is a fabulous magnolia perfume formula by Gerhardt.

**Magnolia Perfume**

| Ingredient | Parts |
|---|---|
| Bergamot (terpeneless) | 220 |
| Lime (terpeneless) | 110 |
| Lemon (terpeneless) | 110 |
| HDXC | 150 |
| Nerolidol | 20 |
| Linalool | 40 |
| Linalyl acetate | 10 |
| Dimethyl anthranilate | 20 |
| Rose (synthetic) | 80 |
| Ylang (high grade) | 50 |
| Jasmine aldehyde | 50 |
| Benzyl acetate | 40 |
| Cinnamyl formate | 10 |
| Citral | 1 |
| Methyl heptine carbonate | 20 |
| Farnesol | 10 |
| Musk ambrettolide | 10 |
| Tolu balsam | 10 |
| Civet tinc. | 20 |
| Indole | 10 |
| Petitgrain | 10 |
| Isobutyl benzoate | 10 |
| Ethyl laurate | 10 |

Following is another magnolia that is nice.

**Magnolia by Secondini**

| Ingredient | Parts |
|---|---|
| Cinnamic alcohol | 75 |
| HDXC | 225 |
| Isoeugenol | 5 |
| Jasmine | 150 |
| Lemon | 75 |
| Musk ketone | 30 |
| Neroli | 225 |
| Vanilla | 15 |
| Ylang | 225 |

The magnolia is cultivated heavily in China, and the flower is used for pickling, flavoring rice, and for herbal formulas.

## The Frangipani (Plumeria)

***Plumeria***

The frangipani is so incredible with notes of jasmine, orange, vanilla, lily of the valley, and coconut with spicy undertones such as carnation. There are some frangipani trees that are hybrids that smell quite different in depth and at different times of the day and the year.

Recently I took a survey for just women on their favorite flower scent. Ninety percent of the women said that frangipani was their favorite flower to smell. I was very surprised at this, since usually, vanilla is number one, especially with women from the mainland USA; however, most of the customers were local people of Maui, and I just assume they are getting tired of jasmine. This is what I have found from interviewing thousands of women when it comes to marketing perfumes. For one, women are very fickle as Morris the Cat when it comes to perfumes and what they like; one day it is jasmine, the next day it is plumeria, etc. Also, it seems they are affected by not only their ovulating cycles but also by their menstrual cycles. So I have pretty much given up on marketing and interrogating women on what they like; now I just concentrate on what I like. Other than this observation, I will not be discussing much more on marketing other than that I do have to make a comment or two on the absolutely ridiculous recorded survey on what odor turned women and men on. The odors that turned men on supposedly were the smell of doughnuts, pumpkin, and lavender. Now, OK, lavender I can wrap my mind around; but come on, doughnuts and pumpkin giving guys erections? I have a really hard time with that one, and the men I interviewed regarding this just cracked up laughing. As far as women, who, like I said earlier, are as fickle as Morris the Cat, their favorite smells were vanilla and cucumber. I can totally understand vanilla, although it does nothing for me; and if it is authentic bourbon vanilla, I just want to eat it. And perhaps they were getting the cucumber confused with the phallic symbol that it represents, but I have never heard of the smell of cucumber causing sexual excitement for a woman (unless the smell reminded her of a sexual experience with a cucumber), whereas the shape, definitely. Furthermore, the women that I interviewed regarding this survey also cracked up laughing. If anything, this kind of marketing is humorous and more than likely a foxy marketing ruse to keep another perfume company from planning a new wonderful perfume launch. Unfortunately these kinds of ruthless childish tactics exist even in the perfume world.

Here are two basic frangipani formulas that are very simple and inexpensive to make. There are three other formulas by Secondini, but they are not worthy of mention.

**Frangipani Imitation No. 1 by Secondini**

|  |  |
|---|---|
| Peru balsam | 30 |
| Bergamot | 120 |
| Carnation (syn.) | 120 |
| Cinnamic alcohol | 20 |
| Coumarin | 30 |
| Lemon oil (terpeneless) | 120 |
| Muguet (syn.) | 130 |
| Neroli | 130 |
| Terpineol | 180 |
| Tuberose (syn.) | 130 |

**Frangipani (Plumeria) by Poucher**

|  |  |
|---|---|
| Jasmine (syn.) | 300 |
| Terpineol | 100 |
| Muguet (syn.) | 80 |
| Rose (syn.) | 200 |
| Aurantiol | 200 |
| Incense | 10 |
| Ben | 30 |
| Vetiver | 10 |
| Orris (syn.) | 20 |
| Coumarin | 20 |
| Musk (your favorite) | 30 |

I have both a Plumeria perfume solid and a single-note Parfum Frangipani on the market.

## The Stargazer Lily

***Stargazer Lily***

Here is one of my favorite flowers, not only to smell but also to look at. I love to arrange these in a vase just to fill my rooms full of their wonderful fragrance and to catch a glance while I am at my perfume organ. They last for days and keep opening just as an old one dies, so you have a continual supply of their luscious scent. When I smell this cut flower, I get whiffs of vanilla, leather, jasmine, ylang, and honey for the bouquet; and when plucking a petal, where the leaf comes off the stem, there is a wonderful smell of fresh grass. As I move closer to the tip of the petal, right in the middle, there is a slight touch of carnation, or maybe a little cinnamon. Tearing into the leaf, there is a watery smell that reminds me of a flowing stream of fresh water, where there are water hyacinths growing alongside lotus blossoms. I smell an orchid garden in the trees hanging over the water's edge. I smell new-mown hay in the leaf that is still wet from just being mowed, and the stamen and pistil smell like rain. There is also a smell overall and in general; I have never smelled this before, and it is quite indescribable.

There are about a hundred species of lilies, and the oriental lilies are the largest, most fragrant of their genus. The stargazer lily is a cross between the oriental lily and the Asiatic, which is smaller and unscented. The main constituents of the Madonna lily, *Lilium candidum*, are p-cresol, phenyl ethyl alcohol, linalool, and terpineol. I do not know the main constituents of the stargazer lily and was unable to obtain any, and I am not sure if any research is being done on it at the present time. The lily is not really a popular odor per se, but I love it, especially the tiger lily. According to Cerbelaud's odor classification, lily is listed under the tuberose-narcissus group, and I would probably have to agree with that.

***Asiatic Lily***

Following is a simple lily base that is quite nice and is inexpensive to make.

**Lily Base by Secondini**

| Ingredient | Parts |
|---|---|
| HXDC | 400 |
| Linalool | 300 |
| Terpineol | 300 |

## The Tuberose

***Double flowering Tuberose***

The tuberose is not like any other flower, and its scent can sometimes be a little overwhelming or offensive for some people. I, on the other hand, could sleep peacefully in a bed of tuberose while others are, vomiting and some people do after only one whiff of this flower. I love everything about it, especially its overpowering odor. It is interesting how its general overall scent is of course tuberose, but on closer examination of the cut flower with my sniffer (sorry, you won't find this word in the dictionary, but it is slang for nose), I smell jasmine, orange flower, carnation, honey, milk, stephanotis, honeysuckle, a touch of lily of the valley, and a hint of coconut. When I tear a petal from the stem, I detect that same green watery smell that I have spoken of previously. Going farther down the stem, there is an overall fresh, almost-squeaky-clean aroma.

I get such a kick out of hearing what a beautiful so-called tropical flower the tuberose is when it originated in Persia, now Iran, and mysteriously migrated to the Americas and Europe and is now cultivated heavily in China. No matter, as long as we have it, I don't care where it comes from. As long as you live, you will probably never smell real, authentic tuberose, concrete or absolute, unless you make it yourself, and that goes for every other essential oil, concrete or absolute, on the market, especially when they go for thousands of dollars for one kilo, which this one does. Tuberose is usually adulterated with methyl anthranilate (which it already has loads of), ylang, methyl benzoate, methyl para-toluate, methyl salicylate, and many other synthetics. One could also "adulterate" their own perfume, so to speak, and this is how great perfumes are born.

Tuberose is full of methyl anthranilate, at least the enfleuraged oil, as opposed to the flower oil obtained by volatile solvent extraction, which is probably why I like it so much. The other main constituents that we know of are nerol and its acetate, eugenol, geraniol, farnesol, methyl salicylate, benzyl benzoate, methyl benzoate, butyric, acid, phenylacetic acid, and tuberlactone.

I have a Tuberose solid on the market named Tuberose Narcose. It is very strong and not for the fainthearted. I also have a single-note Tuberose parfum on the market, but it's not as strong as the solid.

## The Gardenia

I am so fortunate to have a luscious gardenia plant in my garden, and I am constantly looking to see if there is a blossom to bring in to sit on my perfume table. When I smell the gardenia blossom *August beauty*, I smell tuberose whipped with vanilla ice cream, sprinkled with jasmine, iris, ylang, ambrette, carnation, sandalwood, and neroli, with a subtle hint of coconut. Moving down the stem, there is a sharp green note, and I do not smell the watery aroma that I normally smell in most flowers. Overall, it is a wonderful fragrance, and one that I find very difficult to describe.

Supposedly, the main odor of the gardenia is associated with the raw material styrallyl acetate, but I disagree, as I do not think the gardenia flower smells anything like this ingredient, and it could possibly be the reason there are not any incredible gardenia perfumes on the market. Perfumers are taught to use this material when compounding a gardenia. I find that the gardenia is one of the most challenging perfumes to make. Although the perfume made by Annick Goutal named, Gardenia Passion, comes close, it is still not true gardenia. Chanel sporadically releases a gardenia perfume whenever the mood strikes them, but it is never a big seller.

## The Geranium

Geraniums are so amazing insomuch that you can even eat them, bake with them, make ice cream, etc., and I am all for that. Just make sure they are organic before you eat them. There are so many hybrids now, including a chocolate one. The oil is usually inexpensive compared to other oils, except for the bourbon type and the zdravetz. Even more, some great perfumes have been concocted with geranium, and without it, they could have never made it to market.

***Hybrid Geranium***

When I first smell the leaf, there is a burnt odor like the odor of charred sharp green grass. When I tear the leaf apart, there is the same odor now with some fresh water added to it. When I rub the leaf between my fingers, the charred smell disappears, and there is a very strong odor of wheatgrass that is slightly nauseating. The blossom is a little like the leaf, but with rose, honey, and mint. This little flower fascinates me.

Geranium blends well with the rose, of course, and it is used to adulterate rose oil; more than likely, all rose formulas are made with some geranium. The geranium has a similar acid called geranic that is also in the rose, and the other main constituents are geraniol, citronellol, linalool, and eugenol—all mentioned also in the rose. Geranium also mixes well with lavender, clary sage, cedar wood, bergamot, jasmine, lime, neroli, grapefruit, angelica, sandalwood, petitgrain, and most other herbs.

***Hybrid Geranium***

Following is a synthetic geranium.

**Synthetic Geranium by Secondini**

| | |
|---|---|
| Citronellol | 250 |
| Diphenyl oxide | 100 |
| Geraniol | 300 |
| PEA | 175 |
| Rosewood | 175 |

When compounding a synthetic geranium base, you could add to the above trace amounts of eugenol, linalool, along with some rose oxide; and geranium mixed with rosewood at a 1:1 ratio is simply divine.

## Rose

***Rose***

Once upon a time, I had a friend by the name of Tushar Shantam in Santa Fe, New Mexico. We were walking upon the most sacred grounds of the town of Santa Fe, one of the most beautiful and at the same time haunting places that I have ever been or lived in. It is really a graveyard for all of the old Indian spirits and probably should have never been inhabited by humans because the Indians' souls seem to never be at rest there. Anyway, I believe it was the night of my thirtieth birthday, and we had the most delicious bottle of Châteauneuf du Pape from France that was from my birth year. There was an eerie fog covering the grounds like dry ice upon a stage. It was so surreal. We went past an old church that had old roses growing along the fence, and my friend, being the romantic that he was, reached out and grabbed one and handed it to me and said, "Bobbie, how many inhalations of this rose does it take to know its true essence?" I grabbed it into my hand, and without a second thought, I said, "Just one, Tushar." And then I held it to my nose and took the most incredibly long breath, as though it was my last and I was underwater without oxygen and I would never be able to take another. This is the last breath that I want to take before I expire from this world. A perfumed breath. A breath that is my last has to be orgasmic, exalting, powerful, and a statement that I am leaving this realm. This is the feeling and the vibration that I get when I smell the rose. This is what smelling a rose means to me.

Tushar looked at me with such amazement. "Wow," he said, "I thought it would take a few hundred." I was not trying to be romantic, poetic, or funny, as I meant it in all seriousness. When I smell a rose, it only takes one inhalation for me to know all of its essence. When I smell something, I do it with all of my heart and soul; otherwise, the inhalation is wasted. When someone smells me, I expect the same. I want to be smelled like they have never smelled before, and I want them to know all of my essence in one inhalation. When I embrace another human being, I always smell them. I have always done this and have always been conscious of myself doing it.

The rose is the queen of all flowers. She is yin. There are not many perfume formulas made without her. Most of the great perfumes of the past not only had rose, but at least some component of it. Patchouli is her king, very yang, and they wore born to be together. All of my rose perfume formulas are ultimately made with both authentic rose and patchouli. To my knowledge, there is no synthetic patchouli. It is always better the older it gets. It ages like a fine wine, and so whatever perfume formula one makes with it, keep this in mind and allow this concoction to sit longer and brew. I mention in my chapter on making perfumes about letting them sit at least three months, but allow another three months when using these ingredients. Actually, a year is even better.

There are so many roses that I am not sure exactly how many, and I doubt if anyone else knows either. Roses are constantly being hybridized, and they are actually discarded if they don't sell either for the perfume business or the cut flower industry. I am a little saddened by the cut flower industry breeding roses without an odor, but they apparently have to sacrifice it for a longer shelf life. Interestingly, for myself, I personally don't care about the shelf life, and I would much rather have an amazing odoriferous rose emit its heavenly scent for one day than I would like to look at it in a vase for ten days. It is actually hard now to tell some fake roses from the real thing while they are in a vase, and they don't have a smell anyway, so what is the point?

Following is a single-note rose formula I made up a few years ago that is quite nice and has a three—to four-hour evaporation rate on the skin.

**Single Note Rose #1 by Bobbie Kelley**

Jasmine fixative .......... 100
Rose otto .......... 50
Patchouli, aged .......... 20
Bourbon geranium .......... 400

Some natural ingredients that are wonderful in rose compounds, are, of course, patchouli, sandalwood, guaiacwood, benzoin, cedar, clove, frankincense, jasmine, orris, vetiver, geranium, beeswax, etc. Some great synthetics are geraniol, citronellol, nerol, phenyl ethyl alcohol, geranyl acetate, citronellyl acetate, ionones, geranyl formate, phenyl ethyl formate, aldehydes, diphenyl oxide, farnesol, linalool, benzyl acetate, rose oxide, etc. I want to mention the damascones and how amazing they are at transforming rose formulas into something really mind-blowing. The company Ferminich offers these now that they are no longer held captive, and the last time I checked on prices, here was the quote for a kilo of Delta damascone: US $308. Here is a quote for a kilo of Gamma damascone: US $1,039. The former is a floral fruity rose note, and the latter is more of a floral green rose note. One of the best geraniols on the market is by a company named Millennium, and the product is called Geraniol Supra. It is superb, and the last time I checked on the price, it was at US $1,000 per kilo. You get what you pay for. If I were to "adulterate" rose oil, this is what I would use; actually, the grammatically correct word is *sophistication*. I will reiterate, as mentioned earlier in this chapter, this is how great perfumes are born.

I know that the raw material citronellal has gotten a bad rap over the years, but I think it can be wonderful in a rose formula. Of course, it can ruin a rose compound, so use it with caution, and only use it in infinitesimal amounts. It is really used more for soapmaking.

## Acacia

The acacia is one of my favorite flowers, and it is probably one of the most versatile of all plants, being used for teas, tanning, chewing gum, alcoholic beverages, medicine, food, and dyes, and as a host to many other plants. Unfortunately, it is not much used anymore for perfume making because it has been replaced by synthetics. I always try and use real, authentic natural *Acacia farnesiana* or *Mimosa decurrens* instead of synthetics because I believe they are wonderful and add such dimension to a perfume formula. For that matter, I always try and choose as my first choice authentic raw materials in place of synthetics; but unfortunately, this is not always possible. I also believe that the decline of really great perfumes is because of the saturation of so many synthetics, and I will elaborate more on this in another chapter.

***Acacia***

There are many species of the acacia, but one of my favorites is the *Acacia rivalis*, also known as the black or silver wattle from Australia. Several years ago, these were imported and planted on Maui by the Agricultural Department in hopes that they would keep the soil from eroding off the volcano. Instead, what they got was an invasive species (good for me) of trees that grew like weeds and overtook other plants, and laughably, the root system is so weak that one can literally pull a small tree out of the ground with both hands. I have actually done this, so can you imagine this tree ever keeping the dirt from sliding down the mountain? I have watched the topsoil run over these trees like they are microscopic weeds. This is a typical example of the Maui Agricultural Department not doing their homework, and I am always stunned at the ignorance of the department.

The acacia blossoms are very unique, and they do not remind me of any other flower, which is probably why I like them so much. When I first smell the little yellow round ball on the branch, there are rich odors of orange blossom, honey, violet, coumarin, farnesol, and woods. Instead of using forceps on them, because they are so small, I grind them in a mortar; and using a pestle, I crush and grind them for a better whiff. The smell of grass takes over, and as it dries out, there is a smell of old musty sweet bread. In the leaves I detect, a faint odor of costus.

Following is an acacia perfume that I made up a few years ago that is quite nice.

**Acacia by Bobbie Kelley**

| | |
|---|---|
| Mimosa fixative | 300 |
| Acacia, syn. | 300 |
| Ambergris, syn. | 300 |
| Civet tinc. | 10 |

## Mock Orange

***Mock Orange***

I love the smell of mock orange probably because it smells like a mix of the three of my favorite flowers, and they are jasmine, lily of the valley, and orange flower. When I first smell the flower, it smells of the above three flowers mixed with indole. There are also undertones of honeysuckle and gardenia. When I squish the petals in my fingers, there is a smoky green odor that turns into grass.

Following is a lovely formula for mock orange by Gattefosse.

**Mock Orange**

| | |
|---|---|
| Jasmine, artificial | 500 |
| Lily of the valley, artificial | 365 |
| MNK | 25 |
| Methyl ionone | 50 |
| Orange flower petal essence | 50 |
| C 9 – C 10 Aldehyde | 10 |

**Mock Orange by Poucher**

| | |
|---|---|
| Terpineol | 250 |
| Linalool | 150 |
| HDXC | 200 |
| Isobutyl benzoate | 10 |
| Methyl anthranilate | 30 |
| Benzyl acetate | 50 |
| Linalyl acetate | 100 |
| PEA | 150 |
| Vanillin | 10 |
| Indole, 10% | 5 |
| Decyl aldehyde, 10% | 5 |
| Methyl phenyl acetaldehyde | 40 |

**Syringa, Imitation No. 2 by Secondini**

| | |
|---|---|
| Acetophenone | 150 |
| Amyl cinnamic aldehyde | 200 |
| Anisyl formate | 150 |
| HDXC | 150 |
| Isobutyl anthranilate | 50 |
| Isobutyl benzoate | 50 |
| Linalool | 200 |
| Methyl ionone | 50 |

***Mock Orange***

# CHAPTER 6

# Perfumes and the Making Of / Miscellaneous Perfumes

***Paragon Perfumes Mimosa Parfum with authentic mimosa, rose, ylang ylang bergamot and hay. Only 72 bottles made and signed by the Creator/Author***

There is a phenomenon in Chinese medicine called the *running piglet,* and it is a chilling sensation that runs very fast up and/or down the spine. It is quite different than what is sometimes called cold chills. It is also described in ancient Taoist texts as the "coiling of the serpent along the spine." It is also called our qi (pronounced "chee," like you are going to say the word "cheese," but take off the *se* at the end) or our life force as it runs throughout our bodies, especially at times during intense orgasms. This is the experience that I want when I am smelling a raw material to put in a perfume formula when I am creating, and it has happened to me on several different occasions, especially after smelling pure distilled essential oils fresh from my still.

Since I take my sniffing very seriously when I am creating a formula, I will tell you how I prepare myself for it. You could say that it is my own little ritual. First of all, I do not shower for at least twelve hours before I start a perfume creation, nor do I apply any makeup, lotions, deodorant, oils, or perfume during this time. I make sure that my perfumery is well ventilated, and that the temperature and humidity are perfect for my nose and me. I usually want the air conditioner set at exactly sixty-nine degrees Fahrenheit even during the winter months in Hawaii. I turn my cell phone off, and put a Do Not Disturb sign on my door. I do not eat for at least two hours before (the longer the better), and I especially do not consume any dairy products for twenty-four hours (the longer the better) before I start the process. I also take 1,000 mg of vitamin C (nonacidic form) every hour at least two hours ahead because it is an antihistamine. The nasal passages become acutely keen, and then I am ready to start making a perfume formula, or to smell my creations after they have been marinating on the shelf. This is always a very exciting process, and I always look forward to this moment. I want my perfume to be a sensual experience, realizing its essence using my sense of smell and my extra sensory perception, also called e.s.p., meaning I want it to take me on an olfactory journey. See chapter ten.

There are a few different ways of making perfumes; however, I am going to tell you the way I usually (not always) start a perfume formula and then build from there. First, I have an idea of what I want to make, either by a perfume family such as, say, citrus or green, or by a single-note flower, or by something classical, etc. I then think of a fixative or base notes for the formula, usually thinking of raw materials with a high molecular weight for a long evaporation rate, such as balsams, ambergris, musk, some aldehydes, concretes, absolutes, etc. If it were going to be a light formula or cologne, I would not be thinking so much of heavy balsams or musk. I normally like a perfume fixed to the skin; otherwise, I am not much interested in a fleeting perfume. I like something long-lasting so I can savor every nuance about it. Usually, the percentage for the base is about 20–30% of my formula; but then again, I don't want to count because making perfume is not about math. It is about creating magic.

From here, I go to "hearts," or middle notes, such as flower oils, flowery synthetics, and some herbs. The molecular weights can vary depending on how I want the dry out to be. Usually, but not always, included in most of my formulas are the basics, such as rose, ylang, and muguet. Both rose and ylang are great modifiers and can just about smooth or round out any perfume formula. Another is ambrettolide, and I doubt if there is any perfumery in the world without these basic ingredients. Otherwise, the list seems almost infinite for heart notes, and the percentage of these raw materials in my formulas are usually (not always) around 30–60%.

Lastly, I work on top notes, and it could be years before I even get to this point. You could actually skip top notes, but they do add something playful to the final formula. I love citrus top notes, linalool, fresh synthetics, spices, etc. And then of course, some perfumers start with the top notes and then work down to the bottom, going in the opposite direction. The percentage of ingredients that I usually use (not always) is between 10 and 30%. I work with both "open" formulas and with accords.

I have found the best alcohol to use for dilution is grape alcohol, 190-proof ethyl alcohol, to be precise. You can use other alcohols, but only use them as an ingredient, and only as a small percentage to give a little lift or to add something different, but I do not use them as a dilution for the entire perfume.

Now in my opinion, the very best perfume bases that were ever recorded in perfume history were by a man named Rene-Maurice Gattefosse. All of his bases are excellent. The book that I am referring to was first published in the United States in 1959 and also contains some wonderful fixatives and formulas. As simple as some are, they are also profound and quite magical. This book is also out of print and very difficult to find. It took me five years to find a copy, and now I have two. Sheer luck, I presume. There are a few formulas in his book that are quite odd, and I have found myself wondering what Rene must have been thinking. I have made almost all of the formulas in his book and have updated them using the new latest raw materials that I am sure he would have used if he could have procured them. It has been quite the journey. See the chapter devoted to bases and fixatives.

Following is a list of *all-natural*, meaning no synthetics to my knowledge, raw materials that mix well together, whether they are fixatives, hearts, or tops.

Rose and jasmine
Rose and patchouli
Rose, patchouli, and myrrh
Ylang, jasmine, and violet
Rose, neroli, and patchouli
Myrrh, rosewood, and cognac
Vanilla, tonka, and heliotrope
Vanilla, heliotrope, and almond
Sandalwood, amber, and jasmine
Sandalwood, frank, and cassia
Vanilla, Peru balsam, and oakmoss
Myrrh, rose hip oil, and tea rose
Ylang, black pepper, and cognac
Oakmoss, vanilla, tuberose, and Peru balsam
Musk, neroli, and jasmine
Kopi Luwak coffee, ambrette, and vanilla
Patchouli and oakmoss
Jasmine, patch, and sandalwood
Amber, champak, and vanilla
Honeysuckle, lavender, and lantana
Ylang, mandarin, and cardamom
Tolu balsam, vanilla, and horehound
Rose, jasmine, and vetiver
Neroli, cocoa, and amber
Cocoa, lotus, and neroli
Peru balsam, patch, and cocoa
Jasmine, blood orange, and cocoa
Cognac, cocoa, and vanilla (very sweet)
Cocoa, rose, and patch
Patch, carnation, and jaz
Saffron and cocoa
Linden blossom and amber
Bergamot and lavender
Lemon and orange
Verbena and bergamot
Rosewood and lemon
Neroli and orange
Verbena and neroli
Bergamot and thyme
Cognac and myrrh
Tea rose and myrrh
Cypress and ginger
Rosewood, cognac, and myrrh
Oakmoss and Tonka bean
Fir, spearmint, and lime
Jaz, blood orange, and pink grapefruit
Tuberose, jaz, and champak
Anise, hyssop, and vanilla (licorice)
Patch, oakmoss, and vetiver (wood)
Patch, carnation, jaz, and cypress
Green tea, rose, coriander, and pink grapefruit
Oakmoss, jaz, rose, cassie, and orange blossom
Bergamot, lemon, orange, petitgrain, and lavender

The list goes on and on.

Here are some perfume formulas that are quite nice, and although I don't have names for them, I made these simply for fun, and they are only in here as examples of how to build a perfume. All of the ingredients were authentic, unless stated otherwise. My belief is to keep ingredients to a minimum if possible, realizing, of course, that it is much easier said than done.

**Parfum #1 by Bobbie Kelley**

|  |  |
|---|---|
| Sandalwood (Mysore) | 50 |
| Patchouli | 50 |
| Jasmine (synthetic) | 50 |
| Narcissus (syn.) | 15 |
| Ylang (high grade) | 10 |
| Mandarin | 10 |
| Cardamom | 10 |
| Tuberose (syn.) | 100 |
| Orris root | 10 |
| Oakmoss | 10 |

This formula has some very expensive raw materials in it, so if you want to be like most large perfume companies, you could cut costs by replacing with synthetics instead.

**Parfum #2 by Bobbie Kelley**

|  |  |
|---|---|
| Frank | 20 |
| Sambac jasmine | 30 |
| Sandalwood | 50 |
| Vanilla abs. | 10 |
| Heliotrope (synthetic) | 40 |
| Concrete orris | 50 |
| Jasmine grandiflora | 10 |
| Orange blossom abs. | 40 |
| Mimosa abs. | 80 |
| White lily | 80 |
| Ylang (high grade) | 60 |
| Cassia bark | 10 |
| Sweet pea (syn.) | 10 |
| Lily of the valley (syn.) | 100 |
| Rose otto from Turkey | 10 |
| Jonquille abs. | 10 |

All the materials used in this formula were all-natural, meaning no synthetics were used.

**Parfum #3 by Bobbie Kelley**

| | |
|---|---|
| Vanilla | 100 |
| Amber | 50 |
| Patch | 20 |
| Oakmoss | 20 |
| Vetiver | 20 |
| Clove | 20 |
| Incense | 50 |
| Jasmine | 50 |
| Ylang | 100 |
| Rose abs. | 50 |
| Cassia bark | 10 |
| Cinnamon leaf | 20 |
| Peach leaf abs. | 20 |
| Bergamot | 50 |
| Blood orange | 100 |

Again, all of the materials used were authentic.

**Parfum #4 by Bobbie Kelley**

| | |
|---|---|
| Oud | 50 |
| Sandalwood (Mysore) | 50 |
| Jasmine | 100 |
| Sambac jasmine | 10 |
| Rose | 50 |
| Rose abs. | 10 |
| Tuberose | 50 |
| Ylang (high grade or so-called extra) | 80 |
| Orris root | 20 |

**Parfum #6 by Bobbie Kelley**

| | |
|---|---|
| Sandalwood (Mysore) | 150 |
| Patchouli | 50 |
| Fir balsam | 50 |
| Lavender | 100 |
| Coriander | 20 |
| Juniper berry | 50 |
| Lemon verbena | 20 |
| Green lemon from South Africa | 20 |

As above, so below, the ingredients are authentic unless otherwise stated.

**Parfum #7 by Bobbie Kelley**

Sandalwood (Mysore) ........................................ 90
Oakmoss ........................................ 10
Amber ........................................ 100
Jasmine ........................................ 20
Rose ........................................ 20
Tuberose ........................................ 20
Narcissus (syn.) ........................................ 50
Carnation abs. ........................................ 10
Honeysuckle abs. ........................................ 10
Neroli ........................................ 30
Hyacinth (syn.) ........................................ 50
Ylang (extra)........................................ 100
Lilac (syn.) ........................................ 50

All of the ingredients are authentic unless otherwise stated.

**Parfum #8 by Bobbie Kelley**

Sandal (Mysore) ........................................ 70
Cedarwood from the Himalayan Mountains ........................................ 70
Amber ........................................ 20
Vetiver ........................................ 10
Patchouli ........................................ 10
Oakmoss ........................................ 10
Ylang, extra ........................................ 70
Neroli ........................................ 10
Rose ........................................ 10
Jasmine ........................................ 10
Orris ........................................ 10
Tuberose ........................................ 70
Carnation abs. ........................................ 10
Honeysuckle (syn.) ........................................ 10
Black currant bud ........................................ 10
Green lemon from the Ivory Coast ........................................ 10
Concrete lavender ........................................ 70
Hyacinth (syn.) ........................................ 60
Lily of the valley (syn) ........................................ 40
Orange blossom abs. ........................................ 20

I want to elaborate a little on using tinctures, for they are an integral part of perfume making. Historically, perfumers used nothing but tinctures, especially if they could not afford a still. I am a firm believer in the use of alcohol, as I mentioned before, and I have many tinctures in my laboratory ranging from around 0.1% to 50%. The majority of my tinctures are balsams and resins, but you can macerate just about anything. I always allow my tinctures to sit for at least three months; again, six months to one year is even better.

Since I have mentioned balsams and resins, I would like to tell you that for the first four years of my perfume training, I did nothing but work with resins and balsams, and they are the crème de la crème of perfumery. I feel sad for a perfumer when they are unable to work with them in what I call "the great perfumes" because these concoctions are really dependent on them, and the final creation is really the personification of both the perfume and the perfumer. I never allow anyone else to compound my original formulas.

As far as distillation goes, I use only copper stills, and I specialize in the distillation of the various seaweeds of Hawaii. This is a topic for another book and beyond the scope of this one.

There is a phenomenon in the world, called *mass consciousness*, whereby humans and mammals (possibly all other species) think the same thoughts at the same time. This was not recorded and actually proven until recently by scientists, but I always have thought that people did this. Well, I believe perfumers, especially, being somewhat of a different breed, have done this for quite some time. It is interesting, and perfumers have accused other perfumers of stealing their formulas even before there were ways of recording every little detail in the perfume. It actually happened to me. I will tell you the story and give you my formula because I think it is a fascinating phenomenon! By the way, I have not one competitive bone in my body, as I want all perfumers to prosper! That being said, there are literally billions of perfume formulas hanging out in the cosmos just waiting to be made, so I am not at all afraid to be publishing my formulas, as I ultimately believe in total and absolute abundance.

Back in the year 2004, on a full-moon night, I was inspired by my dog—named Honey Girl—to make a sweet perfume and name it after her. For all that I know, she could have thought of the formula and then gave it to me by telepathy . . . LOL. Remember, inspiration can come from anywhere in the cosmos. I remember being very excited about it, but not really having all of the raw materials that I thought should go into such a formula, I went on a search; and for two years, I collected what I thought would be perfect. So in the summer of 2006, on another full-moon night, with all of my ingredients at my side, I went to work at my perfume organ on Honey Girl.

That autumn, I happened to be in the mall shopping at a large department store, and while at the perfume counter, chitchatting with the perfume girl, who knows me very well, I looked over and said to her, "Oh, something new I see. I should try it on." Lo and behold, it was my Honey Girl, which I happened to be wearing, and I laughed and thought the perfumer and I must be very connected; we would have to meet someday. The perfume was by Ralph Lauren, and it is called Hot.

Here is my version, named Honey Girl.

**Honey Girl by Bobbie Kelley**

|  |  |
|---|---|
| Broom absolute | 10 |
| Woody musk | 50 |
| Dark musk and tonka bean abs. mix (1:1 ratio) | 50 |
| Egyptian sandalwood | 50 |
| Dark amber | 120 |
| Vanilla | 20 |
| Heliotropin | 80 |
| Ant. methyl | 100 |
| Honey abs. | 10 |
| Beeswax abs. | 10 |
| Honey (syn.) | 300 |
| Milk and honey (syn.) | 200 |
| Phenyl ethyl acetate | 120 |
| Rose otto from Turkey | 10 |
| Moroccan rose | 40 |
| PEA | 50 |
| Allyl amyl glucolate | 120 |
| Cinalkex | 10 |
| Chamomile | 10 |
| Dimethyl heptanol | 10 |
| Citronellyl acetate | 10 |
| Citronellol | 10 |
| Gerionol | 10 |

Here is a very nice honey perfume that is a lot cheaper to make, and it is a very simple formula by Poucher, from his second volume named Honey No. 1219. I have modified it a little using my honey compound instead of his.

**Honey No. 1219**

| | |
|---|---|
| Bergamot | 200 |
| Rosewood | 200 |
| Geranium | 100 |
| Honey compound (see below) | 100 |
| Amyl cinnamic aldehyde | 50 |
| Civet tinc. | 10 |
| Peru balsam | 60 |
| Musk | 30 |
| PEA | 150 |
| Terpineol | 100 |

**Honey Compound by Bobbie Kelley**

| | |
|---|---|
| Honey (synthetic) | 300 |
| Citronellol | 120 |
| PEA | 350 |
| Styrax (Perfume grade) | 80 |
| Heliotropin | 100 |
| Honey abs. | 50 |

I would like to add a note regarding compounds, and that is that the most important thing for me in compounding single-note compounds is that I always, if possible, add the absolute of the material. This always adds depth to the formula and makes the perfume smell expensive, because it is. Cheap perfumes never have the real stuff! And you only have to use a small amount.

***Variegated Ivy***

The following is a perfume from the green family that I made up a few years ago, and although they are not big sellers per se, they can be fun to wear and are mostly enjoyed by people who really do not want to make a perfume statement.

**A Green Perfume by Bobbie Kelley**

|  |  |
|---|---|
| Hexyl cinnamic ald. | 100 |
| Violet | 70 |
| Viotril, Quest | 30 |
| Vetiver | 10 |
| Vetiveryl acetate | 10 |
| Sweet pea (synthetic) | 80 |
| Ivy (syn.) | 60 |
| Fern extract | 60 |
| Sweetgrass | 60 |
| Coumarin | 40 |
| Orris | 20 |
| Green leaf | 20 |
| Green tea | 90 |
| Linalyl acetate | 70 |
| Linalool | 10 |
| While lily (syn.) | 80 |
| Fig (syn.) | 60 |
| Rosewood | 40 |
| Limes (terpeneless) | 20 |
| Green mandarin | 30 |
| Peony (syn) | 40 |
| | 1000 |

This formula totaled exactly 1,000, but only by accident.

***Heliotrope***

Here is a fun, fruity take on the flower heliotrope, the blossoms of which I love to smell. They smell like a baked cherry pie with vanilla and almonds. I made this a few years ago.

**Fruity Heliotrope by Bobbie Kelley**

| | |
|---|---|
| Black cherry | 10 |
| Hexyl cinnamic ald. | 25 |
| Ant methyl | 25 |
| Jasmine (syn.) | 40 |
| Violet | 80 |
| Sweet pea | 10 |
| Musk, syn. (your choice) | 70 |
| Raspberry | 30 |
| Plum | 40 |
| Anisic alcohol | 40 |
| Vanillin | 180 |
| HDXC | 10 |
| C18 ald. | 10 |
| Ylang | 40 |
| Coumarin | 40 |
| Geranium | 150 |
| Cassia bark | 50 |
| Heliotropin | 180 |
| PEA | 150 |

Here is a single-note tonka bean that I made a while ago and is interesting; it might be nice in a tobacco pipe.

**Single-Note Tonka Bean by Bobbie Kelley**

|  |  |
|---|---|
| Tonka bean abs. | 100 |
| Vanilla | 20 |
| Benzoin | 50 |
| Styrax (perfume grade) | 50 |
| Heliotropin | 80 |
| Musk, syn. (your choice) | 50 |
| Bergamot | 80 |
| Black pepper | 10 |
| Tangerine | 20 |
| PEACE | 80 |
| Cocoa extract | 10 |
| Amber | 50 |
| C18 | 25 |
| Ant. methyl | 25 |
| Benzyl acetate | 50 |
| Amyl cinnamic ald. | 50 |
| Citronellyl acetate | 60 |
| Cognac, French | 50 |
| Oakmoss | 10 |
| Tolu aldehyde | 10 |
| Coumarin | 40 |
| HDXC and PEA mix (1:1) | 30 |
| Tubereuse SAF (Ferminich) | 10 |
| Tobacco (syn.) | 60 |
| | 1000 |

This single-note tonka bean equaled 1,000 also by accident.

Following is a perfume that smells like bubble gum. This is for external use only.

**Bubble Gum Perfume by Bobbie Kelley**

|  |  |
|---|---|
| Lilac, syn. | 250 |
| Linalool | 250 |
| Honey, syn. | 250 |
| Hyacinth Body BHT | 50 |
| Cinnamic alcohol | 50 |
| Clove | 10 |
| Cassia bark | 10 |
| Jasmine, syn. | 70 |

***Pink Hyacinth***

## Perfumed Solids

Perfumes do not have to be in liquid form with alcohol or water. I am talking here about the old-fashioned perfume solids that I really love. There is just something so classy and nostalgic about carrying around a solid perfume. Even though I do not care for antiques, there is something about the thought of going out to listen to a big band, wearing a flapper dress, carrying a long cigarette holder in one hand and a beaded fringed evening bag in the other, with just a tube of lipstick and my favorite solid perfume. I just know that the solid is not going to break and get all over my handmade beaded purse, or worse, on my $8,000 designer dress. It is going to last for years and years (unless I wear it all first), and it is going to smell on me, not smell up the whole, entire nightclub. Most of you young perfumers have no idea at this point what in the world I am talking about, unless you studied it in school, but you old-timers know exactly what I am talking about. I haven't seen perfume solids at perfume counters since the 1980s, at least not in America. And even then, you had to be spending a lot of money or look like you had money. I remember they came in some kind of fancy metal container usually designed with fabulous little flowers. I have seen these old containers (empty and washed, unfortunately) go for quite a large sum on money on eBay, much to my surprise.

So I have brought them back, whether they are in style or not, because I like them. These are made in small amounts because they will last you forever, and they are made with organic and unfiltered beeswax. If you were to mass-market these, of course you would want to use filtered beeswax and calculate the formula differently for larger amounts. Melt the beeswax in a double boiler, and add the oils and synthetics last. And if you really want them to be outrageous, spend a few thousand dollars and buy real, authentic essential oils and or concretes. Do not use petroleum, mineral oil, glycerin, or polysorbate. The musk, heliotropin, and benzyl benzoate are used as fixatives, but you could delete them or use your preferred fixes.

As far as using beeswax and the issue of the bees disappearing, recently I went to a talk here on Maui regarding this heated topic because I am concerned not only about the perfume and cosmetic industry, but also about the cut flower industry, which I am also involved in. My main question was this: Is man/woman interfering with the bees because of the cut flower industry? My main reason for asking this question, which the man looked a little perplexed at, was knowing that bees home in on a flower for several reasons; one is scent. The cut flower industry is only interested in the vase life of a flower and not the scent so in order to have the long life, scent is omitted instead. No scent equals no bee equals no honey equals no food equals death. He did not seem too concerned about this, and his hypothesis was mainly the mite from Africa carried by other bees that know how to clean themselves. His second thought was the change in environment. There always has been, and always will be, moral/ethical issues revolving around most forms of trades. This is definitely one of them.

Here are some of my formulas for perfume solids. All of the formulas use domestic, organic, non-filtered beeswax. I do not use artificial colors of any kind in any of my products.

**Lilac Solid Perfume by Bobbie Kelley**

25 grams beeswax
75 grams lilac (synthetic)
1 drop of civet (synthetic)
10 drops of benzyl benzoate

**Honeysuckle Solid Perfume by Bobbie Kelley**

25 grams beeswax
75 grams honeysuckle (synthetic)
10 drops musk (synthetic)
10 drops heliotropin
10 drops benzyl benzoate

**Lily of the Valley Solid Perfume by Bobbie Kelley**

25 grams beeswax
75 grams lily of the valley (synthetic)
10 drops benzyl benzoate

**Orchid Solid Perfume by Bobbie Kelley**

25 grams beeswax
75 grams synthetic orchid
10 drops benzyl benzoate

***Nobile Dendrobium Orchid***

**Jasmine Solid Perfume by Bobbie Kelley**

25 grams beeswax
75 grams synthetic jasmine
10 drops musk, synthetic
10 drops heliotropin
10 drops benzyl benzoate

***Pink Jasmine***

**Gardenia Perfume Solid by Bobbie Kelley**

25 grams beeswax
75 grams synthetic gardenia
10 drops musk, synthetic
10 drops heliotropin
10 drops benzyl benzoate

**Rose Perfume Solid by Bobbie Kelley**

25 grams beeswax
75 grams synthetic rose
5 drops patchouli
5 drops heliotropin
5 drops musk, synthetic

***Rose***

**Mimosa Perfume Solid by Bobbie Kelley**

25 grams beeswax
75 grams synthetic mimosa
10 drops heliotropin
10 drops benzyl benzoate
10 drops musk, synthetic

**Patchouli Perfume Solid by Bobbie Kelley**

25 grams beeswax
75 grams patchouli (preferably aged)

**Violet Solid Perfume by Bobbie Kelley**

25 grams beeswax
75 grams violet (synthetic)
10 drops benzyl benzoate

**Vanilla Perfume Solid by Bobbie Kelley**

25 grams beeswax
75 grams vanilla compound with bourbon vanilla abs.
1 drop tonka bean absolute
10 drops coumarin
10 drops benzyl benzoate
10 drops musk, synthetic

***African Violets***

People always ask me how long a perfume should last. I tell them it should last forever, at least throughout their lifetime, and who knows how long that will be? The answer is that no one really knows. I have a bottle of Shalimar from 1925, and it takes my breath away every time I open it. My bottle of Cuir de Russie from 1920 is exquisite. The people that were gifted such wonderful gifts almost one hundred years ago are very unlikely going to be alive today to smell from that same bottle of perfume and say, "Yes, that smells exactly like it did one hundred years ago." Or, "Gee, that just doesn't smell as good as it did one hundred years ago." I also have a bottle of perfume that I bought in Germany thirty years ago, and it smells just like a new one off the shelf, perhaps being even a little better, and it has sat in one-hundred-degree temperatures with an extremely high humidity. Unfortunately, I cannot say that the Shalimar smells like a modern one off the shelf, because I do not care for the new change in the formula. So perfume should not really go bad per se, but should live as long as you do, however long that may be. I actually have never had a bottle of perfume spoil, so I am always surprised at this question. I have heard horrible stories about people leaving bottles of perfume in glove compartments of cars, and they have exploded, but it has never happened to me, and I must admit I am guilty of this fault. As a matter of fact, I test all of my perfumes that I make in the glove compartment of my car and subject it to the utmost abuse. I freeze them, bake them, refrigerate them, and throw them around. They should be able to stand up to a lot of mistreatment. I believe if a perfume does spoil on a shelf that is a comfortable temperature and humidity, not under pressure and out of direct sunlight, there is something wrong with that mix. Remember that perfumes were found in ancient tombs in Egypt, which were anywhere from one thousand to five thousand years ago; however, when they were excavated, they had been covered in sand for just as long, so they were protected from a harsh environment, at least for a while. They may not have been the same mix, and they certainly didn't use synthetics, but the concept is the same. As far as perfumes being stored in dark bottles, I don't believe that really makes a difference as long as you use a good UV filter if you are going to be storing it in a clear bottle. The old bottles of perfumes that I mention above are all in clear bottles.

As far as the decline of great perfumes, I believe it to be from the lack of balance in the overall formula, and I attribute this to there being more synthetics than natural raw materials in the perfume. Perfumes nowadays are almost all synthetics, and there should be at least 50% all-natural or authentic raw materials and 50% synthetics at most. Of course, 100% all-natural materials would be ideal, especially if they are organic, which I predict, in the future, all products will be that way again. As much as I like synthetics, they are way overused.

Bottom line is that when it comes to making perfumes, just remember that it is not rocket science, and we are certainly not doing brain surgery, but one does need to develop their own method in how they go about making them; and the perfumer must be able to use both sides of their brain, meaning using both abstract and concrete ideas for the final outcome. If this is not possible, I would suggest teaming up with another perfumer that is dominant in the side of the brain that one may lack.

***Agapanthus***

# CHAPTER 7

# Fougere / New-Mown Hay / Lavender / Clover

Up to this point, I have not written about families of perfumes, except for Russian leather and florals. So I will write about my favorite, which is the fern family, also known as fougere in French. I think it is so amazing that most people do not know that there is such a family, and they say they have never heard of it. They know there is a floral family, and they know there is an oriental family, but fern? Interesting.

When I wrote earlier about there being no indigenous plants of Hawaii, the fern and the silversword, I believe, are the only ones that would stand a chance in a heated debate amongst well-researched botanists. Does the wind count when it blows spores across the ocean? What about birds defecating seeds of plants into the ocean and the seeds being carried for hundreds of miles, swelling in the salt water, waiting to land ashore to germinate in the ground? It is sort of like trying to answer the question which came first, the chicken or the egg? I don't know, but what I do know is that their smell is wonderful, and ferns are found all over the world, especially in tropical and subtropical places. As much as Hawaii would love to claim that certain plants such as the fern, are indigenous, this has never been proven. The so-called Hawaiians aren't even from Hawaii, and they do not know where they are from. For all we know, they could be from another planet, which is a theory as far as the silversword plant is concerned. Now that we have the internet there has been an explosion of information that botanists and plant enthusiasts have recorded because they have found these same plants (that supposedly have been cultivated for centuries) on the islands of Samoa, Cook and Tonga. The Hawaiians have been claiming these same plants are indigenous to the Hawaiian islands. Remember the Hawaiian Islands are relatively new compared to the rest of the earth and the Hawaiians are a very young group of people. There is still an incredible amount of uncharted territory that has never been stepped upon in the planet, and seeds could have come from anywhere. I do understand the ego and the sense of belonging, but as far as history of the fern is concerned, this new information does need careful consideration and further study.

Most of the previous books regarding the making of a fern perfume mention that the fern note was originally made as a *fantasy* concoction because there was no such thing as a scented fern. It is time to banish this old school of thought and realize that some ferns do have a pleasant odor. And this could be the reason why there are not many people who know there are fern perfumes on the market.

I can smell a scented fern a mile away; OK, maybe not a mile, but close, and I love to savor their essence. The fern that I am referring to is the *Phymatosorus scolopendria*, also known as *Microsorum scolopendria*, and is in the Polypodiaceae family, and they are not native to Hawaii. This apparently has been proven. There is also a hay-scented fern family called Dennstaedtiaceae, and this fern is the *Microlepia strigosa*. Another fern, although very toxic, is the *Dryopteris filix-mas*. It is used in medicine as a teniafuge and has to be dosed exactly to body weight, as it is lethal. I have never been able to obtain a sample of this fern, but Steffen Arctander, in his book *Perfume and Flavor Materials of Natural*

*Origin*, describes it as being a dark, viscous liquid that has a sweet, woody, somewhat-earthy, very rich, and tenacious odor. Since most ferns are built on coumarin, one of my loves, naturally, I would be drawn to it. There is something so sweet and primal about coumarin. It is very close to smelling like someone's natural, sweet, warm skin.

***Scented Fern***

As far as I am concerned, there are truly not enough fern perfumes for women on the market.

The fougere note is usually built using the raw materials oakmoss, coumarin, and amyl salicylate, or just oakmoss and amyl salicylate.

The following is a list of raw materials that can be used in a fern perfume:

Lavender
Coumarin
Coumarin derivatives
Flouve
Bergamot
Oakmoss
Amyl salicylate
Benzyl salicylate
Methyl salicylate
Rosewood
Rose
Clary
Geranium

Musk
Rosemary
Patchouli
Vetiver
Vanilla
Sandalwood
Ionones
Cistus
Heliotropin
Eugenol
Citral
Ambrettolide
Spices

***Fern Fronds***

Here is a fougere accord that is very nice, and one could build a wonderful fern with this base.

**Fern Accord**

Bergamot ............................................................................................800
Coumarin ............................................................................................50
Lavender ............................................................................................150

Here is a fern perfume that I made a few years ago that has a very long evaporation rate and outlasted all the other ferns in the batch. It stayed on the blotter for one month, but with most of these ingredients, it would. So if you are looking for a fougere that is fleeting, this would not be the one to wear.

**Fern Perfume by Bobbie Kelley**

| | |
|---|---|
| Oakmoss | 10 |
| Coumarin | 300 |
| Vetiver | 10 |
| Vanilla compound | 10 |
| Tonka bean absolute with musk (1:1) | 50 |
| Sandalwood (Mysore) | 40 |
| Patchouli (aged twelve years) | 10 |
| Ambrettolide | 30 |
| Civet tinc. | 10 |
| Mimosa abs. | 10 |
| Tuberose CP | 10 |
| Rose CP | 10 |
| Jasmine CP | 10 |
| Ylang (high grade) | 20 |
| Rose otto | 10 |
| Geranium CP (geranium with zdravetz) | 40 |
| Myrrh | 30 |
| Clary sage | 10 |
| Heliotropin | 20 |
| Linalyl acetate | 50 |
| Cedarwood | 50 |
| Bergamot (syn.) | 30 |
| Lavender CP | 20 |
| Rosewood | 50 |
| C11 ald. (10%) | 10 |

***Plumosa Fern***

This next formula is my favorite fern cologne, and it was thought of by Rene Gattefosse (supposedly); it is easy and cheap to make. I don't know why, but this brew just rocks my world. This one, however, is short-lived.

**Fern Cologne**

| | |
|---|---|
| Oakmoss | 8 |
| Bergamot, synthetic | 400 |
| Coumarin | 300 |
| Lavender | 100 |
| Geranium | 40 |
| Carnation, syn. | 60 |
| Violet, syn. | 92 |
| Amber | 200 |

***Asparagus Fern***

Following are two fougere formulas by Poucher, and the final result is fabulous.

**Fougere #1017**

Acetophenone ........................................................................5
Benzyl acetate ........................................................................100
Rosewood ........................................................................100
Lavender ........................................................................200
Bergamot ........................................................................300
Amyl salicylate ........................................................................20
Clary ........................................................................10
Anisic aldehyde ........................................................................10
Rose otto ........................................................................5
Jasmine abs. ........................................................................30
Civet abs. ........................................................................30
Coumarin ........................................................................70
Ambrettolide ........................................................................300
Oakmoss ........................................................................20
Patchouli ........................................................................30
Santal ........................................................................40
C11 ald. ........................................................................2
Vanillin ........................................................................5
Vetiver ........................................................................2

**Fougere #1018**

Compound #1017 ........................................................................140
Musk ........................................................................40
Tonka resin ........................................................................10
Ambergris (synthetic) ........................................................................10
Alcohol ........................................................................100

Here are some different formulas that I made, taken from the above Fougere #1017, that are interesting and very nice.

**Fougere Variation #1 by Bobbie Kelley**

Jasmine, synthetic ........................................................................100
Rosewood ........................................................................100
Lavender compound ........................................................................200
Bergamot ........................................................................300
Clary ........................................................................10
Anisyl alcohol ........................................................................10
Rose otto ........................................................................10

Jasmine abs. ....................30
Civet tinc. (1%) ....................10
Coumarin ....................70
Ambrettolide ....................30
Oakmoss ....................20
Patchouli ....................30
Sandalwood ....................40
Vanilla CP ....................10
Vetiver ....................20
Musk, synthetic ....................50
Ambergris, syn. ....................10
Tonka bean abs. ....................1
Anisyaldehyde (Schiff base) ....................10

**Fougere Variation #2 by Bobbie Kelley**

Jasmine, synthetic ....................100
Rosewood ....................100
Lavender compound ....................200
Bergamot ....................300
Clary ....................10
Anisyl alcohol ....................10
Rose otto ....................10
Jasmine abs. ....................30
Civet tinc. (1%) ....................10
Coumarin ....................70
Ambrettolide ....................30
Oakmoss ....................20
Patchouli ....................30
Sandalwood ....................40
Vanilla CP ....................10
Vetiver ....................20
MCK (MCK Coeur, Pearl Chem) ....................50
Canthoxal ....................50
Anisaldehyde (Schiff base) ....................10
Tonka bean abs. ....................10
Exaltolide ....................30
Geranium ....................30
Green lemon ....................30
Satsuma tangerine ....................30
Lemonal ....................30
Citronellyl acetate ....................20
Citral ....................1
Ant. methyl ....................70

**Fougere Variation #3 by Bobbie Kelley**

|  |  |
|---|---|
| Jasmine, synthetic | 100 |
| Rosewood | 100 |
| Lavender compound | 200 |
| Bergamot | 300 |
| Clary | 10 |
| Anisyl alcohol | 10 |
| Rose otto | 10 |
| Jasmine abs. | 30 |
| Civet tinc. (1%) | 10 |
| Coumarin | 70 |
| Ambrettolide | 30 |
| Oakmoss | 20 |
| Patchouli | 30 |
| Sandalwood | 40 |
| Vanilla CP | 10 |
| Vetiver | 20 |
| Cedryl acetate 70% | 50 |
| Tonka bean abs. | 10 |
| Galaxolide | 30 |
| PEACE | 50 |
| Estragole | 10 |
| Aurantiol | 30 |
| Geranium | 40 |
| Lemonal | 40 |
| Green lemon | 40 |
| Satsuma tangerine | 40 |
| Citronellyl acetate | 40 |

***Stag Fern***

**Fougere Variation #4 by Bobbie Kelley**

|  |  |
|---|---|
| Jasmine, synthetic | 100 |
| Rosewood | 100 |
| Lavender compound | 200 |
| Bergamot | 300 |
| Clary | 10 |
| Anisyl alcohol | 10 |
| Rose otto | 10 |
| Jasmine abs. | 30 |
| Civet tinc. (1%) | 10 |
| Coumarin | 70 |
| Ambrettolide | 30 |
| Oakmoss | 20 |
| Patchouli | 30 |
| Sandalwood | 40 |
| Vanilla CP | 10 |
| Vetiver | 20 |
| Cedryl methyl ether | 50 |
| Tonka bean abs. | 1 |
| Pine fixative | 50 |
| Habanolide | 40 |
| Juniper berry | 10 |
| Neroli, syn. | 50 |
| Lemon verbena, syn. | 50 |
| Melissa, syn. | 50 |
| Green lemon | 50 |
| Geranium | 50 |
| Lemonal | 50 |
| Citronellyl acetate | 50 |
| Pink grapefruit | 50 |
| Satsuma tangerine | 50 |

**Fougere Variation #5 by Bobbie Kelley**

Jasmine, synthetic ........................................ 100
Rosewood ........................................ 100
Lavender compound ........................................ 200
Bergamot ........................................ 300
Clary ........................................ 10
Anisyl alcohol ........................................ 10
Rose otto ........................................ 10
Jasmine abs. ........................................ 30
Civet tinc. (1%) ........................................ 10
Coumarin ........................................ 70
Ambrettolide ........................................ 30
Oakmoss ........................................ 20
Patchouli ........................................ 30
Sandalwood ........................................ 40
Vanilla CP ........................................ 10
Vetiver ........................................ 20
C11 ald. (10%) ........................................ 10
Andrane ........................................ 50
Tonka bean abs. ........................................ 1
Velvione ........................................ 30
Vetiveryl acetate ........................................ 20
Carnation CP ........................................ 30
Hedione ........................................ 150
Dimethyl anthranilate ........................................ 20
Rose, syn. ........................................ 100
Geranium ........................................ 50
Green lemon ........................................ 50
Lemonal ........................................ 70
Red mandarin ........................................ 120
Castoreum tinc. (10%) ........................................ 10

Here is yet another fougere that I found in a book by Dr. Olindo Secondini. This formula is very nice and has a slight fruity overtone (caused by the C14 aldehyde) that is playful on my nose. It is inexpensive to make.

**Fougere by Secondini**

C14 ald. ....................30
Bergamot ....................75
Labdanum resin ....................70
Lavender ....................70
Spike lavender ....................240
Oakmoss resin ....................50
Vanillin ....................30
Ethyl alcohol ....................225
Distilled water ....................210

***Scented Fern***

I would like to say that besides my books by Gattefosse and Poucher's classic two-volume set, which is the Perfume Bible, the book by Secondini is wonderful, and I am grateful to have it. I don't know if it is still in print, but it is full of wonderful formulas for both fragrances and flavors.

Patchouli and oakmoss go together like two peas in a pod, but one must be careful when adding these raw materials in a fern because too much can turn it into a chypre. I am not saying that this is bad—I mean, who knows, you might make an incredible chypre by accident if you overdose these ingredients, and I would not be at all surprised if someone told me that is how they ended up making a great chypre. Most of the exquisite perfumes sold on the market are so-called accidents, if not all of them. By the way, my motto is, "If unsure of how much to use, always overdose."

Also, clary sage is one of the most amazing raw materials that we have available, thanks to Gattefosse for really bringing this in vogue. You could probably put clary sage in crap, and it would smell wonderful. Clary sage is not only incredible, but I would say too that it is a great modifier, and can be a top note, heart note, and base note as well. The other great modifiers are, of course, rose, ylang, and ambrettolide. I am fortunate to be able to possess a bottle of real, authentic ambrette, and it is one of the closest scents to heaven; there is nothing else in the world like it. It is another one of those raw materials that you could put it in—well, you know the story. Some other great modifiers are phenyl ethyl alcohol, lavender, neroli, cloves, geranyl acetate, aldehydes, most herbs, methyl and ethyl cinnamate, eugenol, gerianiol, citronellol, cinnamyl acetate, citral, etc.

There is something about the smell of fresh-cut grass that is just wonderful, and I have included new-mown hay perfumes under the ferns because they come very close in odor. It is the constituent coumarin that makes them so powerful. I love coumarin and all its derivatives, and there are perfumes on the market that would have never made it without this incredible ingredient. The tonka bean is close to one half coumarin and, therefore, necessary in an expensive fern or new-mown hay to add depth.

**Foin Coupe No. 1046 by Poucher**

|  |  |
|---|---|
| Acetophenone | 70 |
| Benzyl acetate | 70 |
| Linalool | 300 |
| Lavender | 150 |
| Bergamot | 40 |
| Clary sage | 20 |
| Geranium, bourbon | 50 |
| Benzophenone | 50 |
| Coumarin | 200 |
| Musk, synthetic | 20 |
| Oakmoss | 5 |
| Patchouli | 10 |
| Sandalwood | 15 |

**New-Mown Hay No. 1047**

Compound #1046 (above) ............................................................130
Orange flower absolute ............................................................3
Jasmine absolute ............................................................4
Rose abs. ............................................................2
Civet abs. ............................................................1
Tonka resin ............................................................5
Musk tinc. syn. 3% ............................................................25
Alcohol ............................................................830

The following are four different variations on this formula. If you need to make them to total 1,000, just take some of the alcohol out from no. 1047. For example, the extra ingredients in variation #1 totals 120, so subtract this number from 830 to equal 1,000.

**New-Mown Hay Variation #1 by Bobbie Kelley**

Compound #1047 (see above)............................................................1000
Foin coupe absolute tincture (1%) ............................................................10
Synthetic ambergris, 50% ............................................................10
Benzoin tinc. 10% ............................................................50
Vetiver ............................................................10
Vetiveryl acetate ............................................................10
Sweetgrass ............................................................30

**New-Mown Hay Variation #2 by Bobbie Kelley**

Compound #1047 (see above) ............................................................1000
Methyl anthranilate ............................................................50
Ambergris tinc. syn. 50% ............................................................10
Ambrettolide ............................................................30
Vetiver ............................................................10
Vetiveryl acetate ............................................................10
Fresh-cut grass ............................................................20
Foin coupe absolute tinc. (1%) ............................................................10

**New-Mown Hay Variation #3 by Bobbie Kelley**

|  |  |
|---|---|
| Compound #1047 (above) | 1000 |
| Sweetgrass | 120 |
| Bitter almond tinc. 10% | 10 |
| Vetiver | 10 |
| Vetiveryl acetate | 10 |
| Hay abs. tinc. 1% | 10 |

The New-Mown Hay Variation #3 in particular has a nice, clean scent that would be great as soap or a man's deodorant.

**New-Mown Hay Variation #4 by Bobbie Kelley**

|  |  |
|---|---|
| CP #1047 (above) | 1000 |
| Hay essential oil | 10 |
| Clover compound | 20 |
| Vetiver | 10 |
| Vetiveryl acetate | 10 |
| Foin coupe abs. tinc. 1% | 10 |

I want to talk a little of lavender, and even though I could write a whole book on it, I am going to speak of its highlights and the way I feel about it in general. I am also going to dissect it, literally, and try to describe its scent, then speak briefly of the man I think of each and every time I smell or use lavender in a perfume formula. I guess you could say he is still one of my great inspirations.

***French Lavender***

When I first smell the freshly picked French-type lavender, I detect a floral green sweet mint sage–type odor. When I snap the stem in half, it is even more pronounced, but with a pine scent. When I strip the small leaves off the stem, it is the same former odor, but sweeter and softer.

Without trying to make this book into a healing herbal perfume book, which I very well could, and it has been very hard not to, I have to mention at this point the wonderful antiseptic and healing qualities of lavender. It is by far the most wonderful healing herb on the planet, not to mention its use for aromatherapy, which, I believe, has merit in the perfume world; and eventually, I predict, perfumes of the future will be made wholly and solely for healing.

Even though I also did not want to make this a history book about perfumes, I am fascinated by Gattefosse's story where he was horribly scalded on both of his hands and lower arms in his chemistry lab, and he instantly dipped both hands in what he thought at first was cold water, but it was really lavender oil. His hands were not only healed incredibly fast, but it healed without scarring. I truly believe there are no accidents. I understand that he was a chemist/perfumer, but more importantly, Rene was a physician and researcher. From that day forward in his lab, he knew that he had gone beyond perfume making; he had found his calling to research for the sake of healing. What incredible knowledge he had inside of him, a chemist who had a love for scent and healing the human body. After that day, he experimented like any good mad scientist would and injected copious amounts of essential oils into animals of all kinds for the sake of medicine. Having a history of medicine myself and researching for the Food and Drug Administration of the United States, I can relate to this. Even though we did not experiment on animals, and I would have not been involved in that anyway, I wish Gattefosse would have had a chance to experiment on humans as I did. God bless him! When I say that he was a physician, by definition, he/she is one who places his or her hands on someone with the intention of healing. He may not have been licensed, but so what, for the record, he was a physician.

Now back to lavender: its main constituents are borneol, geraniol, lavendulol, linalool, linalyl acetate, cineole, limonene, pinene, and caryophyllene. It mixes well with clary sage, most citrus oils, geranium, most other herbs, jasmine, patchouli, most spices, and pine.

Here is a lavender perfume I made a few years ago that is quite steely and reminds me of the masculine perfume named Chrome, but better. I was not trying to make a copy of Chrome and it was by sheer accident, for lack of a better word, that it turned out this way. Remember, *mass consciousness*.

**Metallic Lavender by Bobbie Kelley**

Lavender ........................................400
Lavender concrete, Bulgarian ........................................100
Lavender absolute ........................................10
Ambrettolide ........................................20
Coumarin ........................................50
Vetiver ........................................10
Lemon, yellow ........................................50
Bergamot ........................................100
Heliotropin ........................................90
Patchouli ........................................10
Citronellol ........................................60
Neroline yara yara ........................................40
Musk, syn. ........................................ 20
Musk composition ........................................300
Tonka bean absolute ........................................80
Linalyl acetate ........................................100
Geranium ........................................100
Citronellyl acetate ........................................60
Amber ........................................50

Consult Gattefosse's book for more information about lavender formulas; there are way too many for me to put in this book at the present time. Perhaps in a second volume or edition?

Next, I will add some clover-type formulas, which I have added along with ferns and lavender due to the coumarin content and the similarities in the formulas. Following is a lavender-like clover.

**Lavender-like Clover**

Synthetic clover ........................................440
Lavender ........................................250
Bergamot ........................................100
Ylang ........................................100
Coumarin ........................................60
Ambrettolide ........................................40
Phytophenoline ........................................10

**Clover Aroma**

|  |  |
| --- | --- |
| Clover, synthetic | 170 |
| Bergamot | 360 |
| Vetiver | 10 |
| Ylang | 160 |
| Clove | 40 |
| Geranium | 60 |
| Coumarin | 100 |
| Vanilla | 90 |
| Phytophenoline | 10 |

***Clover***

**Clover Perfume**

| Ingredient | Parts |
|---|---|
| Clover, synthetic | 350 |
| Linalyl acetate | 80 |
| Linalool supra (Millennium) | 160 |
| Lilac | 50 |
| Lavender | 80 |
| Lavender concrete | 30 |
| Oakmoss | 10 |
| Heliotropin | 30 |
| Coumarin | 100 |
| Patchouli | 10 |
| Synthetic sandalwood | 30 |
| Vanilla | 20 |
| Violet, syn. | 120 |
| Musk, synthetic | 40 |
| Ambrettolide | 30 |
| Phytophenoline | 1 |

I want to mention the preservatives by the company Bio-Botanica, based out of New York, called Phytophenoline, Biopein, and Neopein, since I used the former in the above formula. These are supposedly all-natural and, I believe, have great potential in the world of perfumery and cosmetics. The people working there are wonderful to talk to, and their chemist is very knowledgeable on the art of perfumes; being from Egypt, I believe he has great merit in the world of perfumery.

Biopein consists of these extracts:

*Origanum vulgare* leaf
*Thymus vulgaris*
*Cinnamomum zeylanicum* bark
*Rosmarinus officinalis* leaf
*Lavandula angustifolia* flower
*Hydrastis canadensis*

Neopein consists of these extracts:

*Origanum vulgare* leaf
*Thymus vulgaris*
*Cinnamomum zeylanicum* bark
*Rosmarinus officinalis* leaf
*Lavandula angustifolia* flower
*Hydrastis canadensis*
*Olea europaea* leaf

# CHAPTER 8

# Bases and Fixes

There are probably more bases on the market than there are hybrid roses, but I make most of my own. As far as Schiff bases go, I also make those as well and use them quite frequently. Methyl anthranilate is one of my favorite materials to work with, and I have mixed numerous aldehydes at different proportions with it, and all of them are fascinating. It is certainly not for every perfume, but when it is used at the right dose and for the right family of perfumes, there is nothing else like it. My favorite Schiff base is aurantiol, and I use it in a lot of formulas. It took me years to get used to aldehydes, but they are remarkable and, sometimes, can give astounding results; they are challenging and playful, therefore I really enjoy them. Even though this chapter is not about aldehydes, they are very relative and important to bases. Other aldehydes I really like and find interesting are ortho-methoxycinnamic, Lysmeral (even though expensive, it is better than Lyral and Lilial, in my opinion), cortex aldehyde, cyclamen aldehyde, muguet aldehyde, and both amyl and hexyl cinnamic aldehydes. The Schiff bases increase the tenacity and stability of the aldehyde.

I want to touch a little on fixatives, since they are so important. Again, the best fixatives I have ever found in any book were by Rene-Maurice Gattefosse. The dry out on these are amazing and extremely long-lasting. And if you tweak his formulas just a little, you get something very strange and wonderful, depending on what you use. I must add that in the recent years, "fixing" a perfume has not really been in style, and this could be one of the reasons that there have not been any books (that I know of) published on the art of fixation since 1959, especially using real and authentic ingredients such as resins and balsams. Resins and balsams are messy, but they are worth their weight in gold. It is my belief that the decline of perfumes is because of the overuse of synthetics and not enough resins and balsams; in other words, there is no balance. By all of the consumer reports, most clients wanted something fleeting and not cloying, but I, on the other hand, loved the '60s, because when a lady wore a perfume, you could smell it on her for hours on hours. I also remember it still being on my clothes after a long night out disco dancing in the '70s, and usually with great memories to match. I really love using fixes, bases, and raw materials with a long evaporation rate and a high molecular weight. There could also be some psychology to it, whereas I really want my money's worth, or more bang for my buck; after all, I still am a consumer. Whatever the reason, it is what I prefer.

The following fixatives are very tenacious and are of a very low volatility. As far as fixatives go, besides the all-natural resins, which I believe are the best, I want to mention the synthetic called Fixateur 505 D by Ferminich, which is really wonderful. And there are hundreds of other synthetic fixatives that are beyond the scope of this book.

Following is a base that is simple yet very elegant.

**Rose Base by Bobbie Kelley**

Cedarwood ....................100
Jasmine, authentic ....................10
Geranium ....................300
Clove ....................10
Citronella ....................1
Orris, authentic ....................10
Palmarosa ....................50
Frank ....................50
Vetiver ....................10

**Rose Fixative by Bobbie Kelley**

Incense ....................160
Orris ....................80
Vetiver ....................30
Cedarwood ....................30
Citronella ....................2
Geranium ....................120
Sandalwood ....................50

Citronella has received a bad rap, but don't be afraid to use it; but only dose it in minute or trace amounts. It is used more for candle making. Of course, the quality is important just like any other raw material, so be selective of your supplier.

**Magnolia Base (Poucher)**

Nerol ....................300
HDXC ....................200
Lemon oil (terpeneless) ....................100
Ylang-ylang (high grade) ....................100
Cinnamyl alcohol ....................100
Isoeugenol ....................10
Vanilla abs. ....................10
Jasmine compound ....................130
Musk ketone ....................50

Here is a fixative that I made up that I have enjoyed working with, it's called Moss Mix Fix. I designed it mainly for chypre perfumes, but you can use it with just about anything.

***Unknown Moss***

**Moss Mix Fix by Bobbie Kelley**

Scentenal aldehyde ........................................ ½ oz.
Oakmoss ........................................ ½ oz.
Evernyl tinc. (10%) ........................................ ½ oz.
Pine tree moss abs. tinc. (5%) ........................................ ½ oz.
Cedar moss tinc. (5%) ........................................ ½ oz.
Patchouli (aged eleven years) ........................................ ½ oz.
Dark synthetic musk (or use your favorite musk) ........................................ ¼ oz.
Tonka bean abs. ........................................ ¼ oz.
Styrax resin (perfume grade)........................................ ½ oz.
Dark amber, synthetic ........................................ ½ oz.
Vetiver ........................................ ½ oz.
Ambrettolide ........................................ 1 oz.
Ambergris (syn) tinc. (50%) ........................................ ½ oz.

Patchouli is like a fine cabernet wine, the older it gets, the better it gets.

The following bases and fixatives have been adapted from Gattefosse's classical formulas from his book that I have mentioned before. I have personalized them to my liking.

**Chypre Base by Bobbie Kelley**

Bergamot, terpeneless ....................170
Oakmoss ....................50
Patchouli ....................10
Lavender ....................50
Orange, terpeneless ....................90
Sandalwood ....................50
Cetyver ....................80
Gerionol special ....................110
Coumarin ....................50
Musk ambrette ....................20
Musk, synthetic ....................50
Heliotropin ....................100
Jasmine, synthetic .................... 40

**Wood Base by Bobbie Kelley**

Bergamot, terpeneless ....................250
Rose, authentic ....................50
Sandalwood ....................40
Vetiver ....................110
Cetyver ....................50
Cassia bark ....................40
Storax, perfume grade ....................80
Coumarin ....................70
Exaltolide ....................40
Ambrette ....................20

**Moss Base by Bobbie Kelley**

Bergamot, terpeneless ....................120
Patchouli ....................20
Vetiver ....................100
Linalool Supra, Millennium ....................120
Musk, synthetic ....................100
Heliotropin ....................120
Coumarin ....................120
Amber ....................60
Orris base ....................20

**Orange Fixative by Bobbie Kelley**

Benzoin ..........300
Tolu ..........200
Vanilla absolute ..........100
Myrrh ..........150
MNK ..........50
Ant methyl ..........50
Yara yara ..........50
Clary sage ..........30
C10 aldehyde ..........10
Phenyl ethyl phenylacetate .......... 60

**Violet Fixative by Bobbie Kelley**

Orris ..........200
Ambrette ..........100
Myrrh ..........120
Orris concrete ..........70
Coumarin ..........50
Vanilla absolute ..........20
Violet leaf absolute ..........50

**Vanilla Fixative by Bobbie Kelley**

Tonka bean ..........200
Vanilla absolute ..........300
Tolu ..........100
Cassia ..........80
Ethyl vanillin ..........100
Heliotropin ..........40
Coumarin ..........20
Clove bud ..........60

# CHAPTER 9
# Woods / Musk / Ambers / Chypre

***Eucalyptus Tree***

I love the woods, and there is a part of me that finds all trees sacred. When I was about ten years of age, my father purchased a rather large parcel of land in the hills of Kentucky that was saturated like a jungle with the most beautiful and majestic old oak and elm trees, among many others. One day my father decided to sell some of the old oak trees to a well-known, high-class furniture-making company to help clear the land. They would do the cutting and hauling, so it seemed to be a win-win situation. Of course, they wanted the largest oak trees, which were also the oldest, because they could make bigger pieces of furniture. The day came for the men to cut them down, and I can remember my family saying a blessing. There were about six of us at the time, and we had gathered in a circle around the first tree to be cut; and clasping our hands, we said a silent prayer. My father had said that the Indians did this if they needed wood for a fire, and that we should do it too. I will never forget the sound of that chain saw that day or the tears in my father's eyes as the first cut was made. He turned his head and was unable to look at the event as it ensued. I had only seen my father cry one other time in my life, and it was when I was much older. I watched him light up one cigarette after another as I observed the intense mental anguish consuming his mind and body as he struggled with how he would contain his emotions. As soon as my father heard the tree fall, he walked quietly over to the men and asked them nicely to leave the property. They could take the tree, but no more, and to never return or ask to cut any more trees. He never spoke of it again, and neither did the rest of the family. Apparently, by the rings of the tree, it was around two hundred years old.

***A very large old Camphor Tree***

As much as I like synthetics, there is nothing like authentic woods, and I probably love the woods more than I do flowers; however, there has been such controversy over using real wood oils. I am always perplexed at people who advocate "Save the planet" and "Go green," and yet at the same time, they want a perfume with, let's say, authentic sandalwood. They get angry when the sandalwood trees are chopped down, yet they also get angry with the chemists who have worked so hard trying to replace the sandalwood with synthetics to mimic sandalwood. Yet these same people go and chop down a tree for Christmas, watch it light up for two weeks, and then go throw it in the dump. I don't get it; am I missing something?

While I am on the subject of sandalwood, I do want to address the issue of what I call lack in the minds of most people—most likely due to watching way too much TV, fear, environmental factors, and brainwashing. I would like for more people to, instead of complaining, "visualize" more sandalwood trees, or better yet, go out and plant one. Actually, there are thousands of acres of planted sandalwood trees in Australia, and people are in the process of planting more. Perhaps there could be some kind of trade or compromise; let's say for every case of perfume sold that contains real, authentic sandalwood, the company could plant a baby sandalwood tree, or perhaps two, or better yet, a whole acre. I know that other companies are doing this, and they are advocating that if you do things like open a new bank account, sign up for a new credit card, buy a new house, etc., they will plant a tree. I am more than willing to do this, and have already done it, because I believe in giving back to Mother Earth. I believe it to be an act of good faith toward mankind/womankind. Or how about this: instead of giving a free gift such as a scarf, handbag or watch when you purchase a bottle of perfume, give with it a baby tree to plant. The tree could have the same theme as the perfume. For example, if it is a magnolia perfume, give away a baby magnolia tree. Gee, I wasn't going to talk about marketing anymore, was I?

Maybe there should be a new law. Instead of going to jail for walking down the street having a glass of wine while you are window-shopping, you would go to jail if you get caught chopping down a tree and not replanting another one. I am being facetious, of course, actually mocking some of the ridiculous laws in America and some other Middle Eastern countries that I have visited.

Back to the woods—so most, if not all, perfumes can benefit from adding a wood or a woody composition. Most of all of the classic perfumes have some woods in them. I have never been able to find a really great sandalwood synthetic replacement, but you can come close with a mixture of some of the products on the market such as Santalidol, Polysantol, Sandela, Indisan, Sandel N, Sandalore, etc.

Following are some wood formulas that are quite nice. These are modified from Gattefosse's woods and are my renditions.

This formula is very interesting in that for the first two months, it smelled like Indian curry, and then it mellowed out during the last thirty days and became very lovely.

**Indian Woods 8 by Bobbie Kelley**

|  |  |
|---|---|
| Sandalwood | 120 |
| Iralis Total 949970, Ferminich | 80 |
| Cypress, blue | 40 |
| Ylang | 30 |
| Geranium | 30 |
| Patchouli | 20 |
| Vetiver | 80 |
| Rose, synthetic | 30 |
| Amber | 30 |
| Vanillin | 10 |
| Coumarin | 30 |
| Ambrettolide | 30 |
| Exaltolide | 30 |
| Cetyver SA, Ferminich | 80 |

*Mango Tree*

*Mangoes*

**Gilded Wood by Bobbie Kelley**

| Ingredient | Parts |
|---|---|
| Cetyver | 100 |
| Iralia | 100 |
| Violet, syn. | 200 |
| Styrax | 100 |
| Heliotropin | 100 |
| Neroli | 20 |
| Ylang extra | 40 |
| Jasmine, syn. | 80 |
| Buddahwood | 20 |
| Ambrettolide | 50 |
| Exaltolide | 50 |
| Fixateur 505D | 50 |

**Palisander Wood by Bobbie Kelley**

| Ingredient | Parts |
|---|---|
| Sandalwood, syn. | 100 |
| Vetiver | 60 |
| Neroli, authentic | 50 |
| Bergamot, syn. | 50 |
| Oud, syn. | 100 |
| Rosewood | 30 |
| Jasmine, syn. | 80 |
| Exaltolide | 50 |
| Anisyl alcohol | 40 |

**Pagodes Wood by Bobbie Kelley**

| Ingredient | Parts |
|---|---|
| Sandalwood, authentic | 150 |
| Cetyver | 100 |
| Vetyrisia | 40 |
| Bergamot | 170 |
| Civet tinc. | 10 |
| Rosewood | 80 |
| Rose otto, authentic | 10 |
| Lavender concrete | 50 |
| Coumarin | 10 |
| Vanilla abs. | 1 |
| Opopanax | 80 |
| Fixateur 505 | 50 |
| Amyris | 50 |

**Havana Wood by Bobbie Kelley**

Guaiacwood ....................50
Sandalwood, authentic ....................50
Teak, syn. ....................50
Buddahwood ....................50
Rosewood ....................50
Amyris ....................50
Violet ....................250
Ylang extra ....................20
Bergamot ....................60
Geranium ....................50
Tolu balsam ....................30
Cistus ....................50
Civet tinc. ....................10
Heliotropin ....................70
Ambrettolide ....................70
Exaltolide ....................50
Incense ....................50
Fixateur 505 ....................50

**Siam Wood by Bobbie Kelley**

Sandalwood, synthetic ....................100
Vetiver ....................60
Teakwood ....................60
Bergamot ....................150
Rose, syn. ....................30
Jasmine abs. ....................10
Coumarin ....................80
Fixateur 505 ....................50

*Jacaranda Tree*

*Jacaranda Blossoms*

**Cedar Pencils by Bobbie Kelley**

| | |
|---|---|
| Cedral methyl ether | 50 |
| Acetophenone | 5 |
| Benzyl acetate | 100 |
| Rosewood | 100 |
| Lavender | 200 |
| Bergamot | 300 |
| Amyl salicylate | 20 |
| Clary | 10 |
| Anisic aldehyde | 10 |
| Rose otto | 5 |
| Jasmine abs. | 30 |
| Civet abs. | 30 |
| Coumarin | 70 |
| Ambrettolide | 300 |
| Oakmoss | 20 |
| Patchouli | 30 |
| Santal | 40 |
| C11 ald. | 2 |
| Vanillin | 5 |
| Vetiver | 2 |

I would like to mention some wonderful synthetic woods that I find interesting and have used in a lot of my formulas. They are Andrane, Vertofix, Timberol, Iso E Super, Cedramber, Bacdanol, Tabacarol, and Spirambrene. Tobacarol is ludicrously expensive, but it can turn a formula from "just OK" to "wow!"

***Cook Pine***

## Now on to musk . . .

I believe if Rene Gattefosse were alive today, he would tell us he would have used copious amounts of exaltolide in his musk formulas. Exaltolide is my favorite synthetic macrocyclic musk, and it is chemically analogous to testosterone, which is in both men and women. It is just that most men have way more testosterone than women. This is one of the reasons that I believe that both men and women are able to wear musk, and it smells completely natural. As far as coed, bi, shared perfumes, I have always believed that all perfumes are for both men and women. I have always worn men's cologne, and still do, and I know several men who wear women's perfume. Whenever I start creating a new perfume formula, I do not think in terms of, "Is this going to be for a male or a female?" I really believe this thwarts the creative process and is one of the reasons perfumers have a hard time with customer requests.

There are some other synthetics besides exaltolide that mimic musk that are quite nice. It is most unfortunate that we have to use them because there is nothing like real musk; however, on the other side of the coin, I am grateful to have them because I would like to be able to save the deer's life. The synthetics worthy of mention are habanolide, galaxolide, celestolide, musk R1, and velvione. As of the time of this writing, a kilo of musk R1 sells for US $550. As far as I am concerned, that is still cheaper than $45,000 a kilo for real musk, and it can save the lives of two to four thousand musk deer.

From a report by TRAFFIC (an international wildlife trade-monitoring network) and WWF (a conservation organization) in 1999 regarding the musk deer, *Moschus moschiferus*, there were approximately 600,000 deer in existence in China, but this was only an estimate. From their ongoing reports, it could be around that amount today. The largest number of musk deer was in Russia and China, and they also were found in Kazakhstan, Kyrgyzstan, Korea, and Mongolia. Only the male musk deer produces musk, and only about 25 grams per year per animal, so it would take the killing of two to four thousand male deer for a kilo of musk, and that is only if the deer has large-enough glands. Even the females and young males are accidentally shot while trying to catch the males that are old enough to produce the musk in their glands. It is possible to obtain the musk from a live deer, but most are killed first, and it is done by sedation after being caught in nets.

All musk deer have been on the endangered species list since 1979, and apparently, there are six or more species of the musk deer. According to this report, up to 1996, France, Germany, and Switzerland were indirectly responsible (whether they killed the deer themselves or not) for the killing of tens of thousands of musk deer as they were the major importers. This figure had gone down from 1996 to 1999, but I don't believe that people really cared for the saving of the animals' lives, or they would have stopped it many years ago. I really believe it to be because it became so expensive to buy the musk; for instance, in Japan, musk is up to $35–$50 per gram, but this is nothing compared to cocaine, which still sells for over $100 per gram. And the price of musk in Europe in the 1990s had reached three to five times that of gold. In 1999, musk was purchased at about $12–$14 per gram in Europe and South Korea. In 1997, there was a recorded shipment of musk from Russia to Germany for a price of approximately $16–$22 per gram.

In Europe, apparently, there is one zoo in Germany that is breeding the Siberian musk deer, and it is called the Leipzig Zoo. There are also zoos in Berlin, Paris, and Italy that have the deer; but the mortality rate is very high, and they do not do well in captivity. I believe there is also a musk deer at the San Diego Zoo in the United States. There are also deer farms set up in China, and in 1989, there were recorded only 2,000 musk deer on these farms in the entire country.

I do not purchase real, authentic musk deer products or any other animal musk products, nor do I use them in my perfume formulas, and I will no longer purchase perfumes made with real, authentic musk. Even if someone told me the musk was from a live deer, I would not buy it. First of all, I would not believe them; and secondly, I don't appreciate the fact that they are scraping the glands of a live animal. This musk is imperative for breeding and propagation of the species. I sort of imagine myself in the place of the musk deer and think what if someone held me down and scraped my scent glands. No way!

Following are some musk formulas that can be used in any perfume formula that are modified from Gattefosse's musk compounds. These are my versions.

**Musk for Extract 1 by Bobbie Kelley**

Musk, your choice ....................70
Exaltolide ....................20
Benzoin ....................20
Labdanum absolute resin ....................50
Ambrettolide ....................20
Tolu balsam ....................30
Sandalwood ....................80
Rose CP with authentic rose ....................280
Bergamot, authentic ....................260
Heliotropin ....................90

**Musk Composition for Extract 2 by Bobbie Kelley**

Musk, your choice ....................70
Grisambrol ....................30
Exaltolide ....................20
Sandalwood, authentic ....................80
Rose compound with authentic rose otto ....................280
Bergamot, authentic ....................100
Clary sage ....................10
Rose absolute ....................20
Rose, artificial ....................260
Heliotropin ....................90
Geranium ....................100
Violet, synthetic ....................90

**Liquid Musk for Extract 3 by Bobbie Kelley**

Musk, your choice ....................200
Ambergris, syn. ....................150
Exaltolide ....................20
Rose, syn. ....................240
Patchouli ....................40
Bergamot, syn. ....................40
Sandalwood ....................20
Jasmine abs. ....................30
Geranium ....................50
Orris concrete ....................20

Cedarwood ....................................................................................20
Labdanum abs. ..............................................................................50
Phenyl ethyl alcohol .......................................................................40
Clary ..........................................................................................10

Next are formulas using the above compositions, and they are examples of what you can do with these. You can use these as bases or fixatives in any perfume, and/or you can also replace your favorite musk compounds where I have used the words "your choice."

**Musk for Extract 1 with Celestolide by Bobbie Kelley**

Musk, your choice .........................................................................70
Exaltolide ....................................................................................20
Benzoin .......................................................................................20
Labdanum absolute resin ...............................................................50
Ambrettolide ................................................................................20
Tolu balsam ..................................................................................30
Sandalwood ..................................................................................80
Rose CP with authentic rose ..........................................................280
Bergamot, authentic ......................................................................260
Heliotropin ...................................................................................90
BOB ...........................................................................................850
Civet, undiluted ............................................................................10
Celestolide ..................................................................................110

**Musk for Extract 1 with Phytoplenolin by Bobbie Kelley**

Musk, your choice .........................................................................70
Exaltolide ....................................................................................20
Benzoin .......................................................................................20
Labdanum absolute resin ...............................................................50
Ambrettolide ................................................................................20
Tolu balsam ..................................................................................30
Sandalwood ..................................................................................80
Rose CP with authentic rose ..........................................................280
Bergamot, authentic ......................................................................260
Heliotropin ...................................................................................90
BOB ...........................................................................................850
Phytophenolin, Bio-Botanica ..........................................................10

Phytophenolin is both a preservative and booster.

### Musk Composition for Extract 2 with Cashmeran by Bobbie Kelley

| | |
|---|---|
| Musk, your choice | 70 |
| Grisambrol | 30 |
| Exaltolide | 20 |
| Sandalwood, authentic | 80 |
| Rose compound with authentic rose otto | 280 |
| Bergamot, authentic | 100 |
| Clary sage | 10 |
| Rose absolute | 20 |
| Rose, artificial | 260 |
| Heliotropin | 90 |
| Geranium | 100 |
| Violet, synthetic | 90 |
| BOB | 850 |
| Cashmeran | 110 |

The above can be used for an oriental-type base or fixative.

### Liquid Musk for Extract 3 with Cashmeran by Bobbie Kelley

| | |
|---|---|
| Musk, your choice | 200 |
| Ambergris, syn. | 150 |
| Exaltolide | 20 |
| Rose, syn. | 240 |
| Patchouli | 40 |
| Bergamot, syn. | 40 |
| Sandalwood | 20 |
| Jasmine abs. | 30 |
| Geranium | 50 |
| Orris concrete | 20 |
| Cedarwood | 20 |
| Labdanum abs. | 50 |
| Phenyl ethyl alcohol | 40 |
| Clary | 10 |
| Civet, undiluted | 10 |
| Cashmeran | 110 |
| BOB | 850 |

**Liquid Musk for Extract 3 with Celestolide and Cashmeran by Bobbie Kelley**

| Ingredient | Parts |
|---|---|
| Musk, your choice | 200 |
| Ambergris, syn. | 150 |
| Exaltolide | 20 |
| Rose, syn. | 240 |
| Patchouli | 40 |
| Bergamot, syn. | 40 |
| Sandalwood | 20 |
| Jasmine abs. | 30 |
| Geranium | 50 |
| Orris concrete | 20 |
| Cedarwood | 20 |
| Labdanum abs. | 50 |
| Phenyl ethyl alcohol | 40 |
| Clary | 10 |
| BOB | 850 |
| Celestolide in PEA | 110 |
| Cashmeran | 110 |
| Civet tinc. | 10 |
| Coumarin | 100 |
| Vetiver | 10 |
| Rosewood | 50 |
| Bergamot, syn. | 300 |
| Cistus | 50 |

***Cistus (Rock Rose)***

Again, the last two can be used in an oriental formula.

**Bobbie's Oriental Base (BOB) by Bobbie Kelley**

Vanilla CP with authentic bourbon vanilla ....................50
Frankincense ....................50
Myrrh ....................50
Ambrettolide ....................50
Ambergris, syn. ....................100
Sandalwood, authentic ....................50
Patchouli ....................50
Opopanax ....................50
Fixateur 505 ....................50
Roseessence 193D ....................100
Jasmine CP w/ jasmine absolute ....................100
Jasmolactone ....................50
C12 ald. lauric (50%) ....................10
Cocoa absolute ....................10
C18 ald. ....................10

The synthetic grisambrol is very versatile and can be used in just about anything such as Russian leathers, mosses, ferns, marines, musk, and especially woods and ambers.

Following is another family of perfumes that is one of my favorites, and that is amber. I am talking about amber perfumes, and not the resin amber, which is a forty-million-year-old fossil, which I also love. Of course, I will not use real ambergris unless I find it myself floating aimlessly in the ocean or washed up onshore. I wouldn't believe someone even if they told me that is where they found it. Sorry. Bottom line is I will not support the slaughtering of a sperm whale just to get his ambergris, and I believe, unfortunately, that there are still people out there in the market of killing sperm whales just to make a fast buck. No matter what they tell me, unless I find the ambergris myself, forget it.

Following is a list of synthetics that can be used when compounding ambers.

Ambrox
Geosmin
Grisalva
Fixateur 404
Alpha-Ambrinol
Ambraketal (also woody)
Etc.

**Amber by Bobbie Kelley**

Labdanum absolute ......................................................................................80
Ambergris, syn. ..........................................................................................80
Ambrettolide ..............................................................................................80
Oakmoss ....................................................................................................30
Vetiver ........................................................................................................80
Rose CP .....................................................................................................30
Violet, syn. ................................................................................................150
Vanillin .......................................................................................................50
Heliotropin ..................................................................................................70
Benzoin resin ............................................................................................120

**Amber Variety by Bobbie Kelley**

Labdanum resin .........................................................................................350
Bourbon vanilla ..........................................................................................150
Tonka bean abs. ...........................................................................................70
Tolu balsam resin .......................................................................................100
Sandalwood ..................................................................................................80
Rose concrete ...............................................................................................50
Jasmine abs. ..................................................................................................30

**Amber Variation #1 by Bobbie Kelley**

Labdanum abs. ..............................................................................................50
Artificial amber ...........................................................................................180
Ambergris, syn. .............................................................................................30
Clary...............................................................................................................30
Musk, your choice .........................................................................................60
Ambrettolide ..................................................................................................60
Benzoin ........................................................................................................230
Rose CP .......................................................................................................160
Jasmine abs. ..................................................................................................10
Heliotropin.....................................................................................................50
Vanillin ..........................................................................................................20

**Amber Second Variation by Bobbie Kelley**

Artificial amber ....................................................................................300
Bergamot ..............................................................................................200
Rose CP ..................................................................................................50
Tolu balsam ...........................................................................................150
Musk, your choice ....................................................................................40
Ambrettolide ............................................................................................20

Ambers beg for flowers to be added to them. Following are two of my formulas that are quite nice.

**Amber Variety with Acacia by Bobbie Kelley**

Labdanum resin .....................................................................................350
Bourbon vanilla .....................................................................................150
Tonka bean abs. ......................................................................................70
Tolu balsam resin ..................................................................................100
Sandalwood .............................................................................................80
Rose concrete ..........................................................................................50
Jasmine abs. ............................................................................................30
Acacia, syn. ...........................................................................................300
Ambergris, syn. ......................................................................................150
Civet, undiluted .......................................................................................20
Ylang .......................................................................................................80
Celestolide w/ PEA ................................................................................100
Cashmeran .............................................................................................100
Mandarin Napoleon .................................................................................80
Pink pepper ..............................................................................................10
Aurantiol ..................................................................................................50
Clary ........................................................................................................10
HDXC ....................................................................................................100
Hedione ..................................................................................................100
Tubereuse SAF .........................................................................................40

***Acacia***

**Amber Second Variation with Violet by Bobbie Kelley**

|  |  |
|---|---|
| Artificial amber | 300 |
| Bergamot | 200 |
| Rose CP | 50 |
| Tolu balsam | 150 |
| Musk, your choice | 40 |
| Ambrettolide | 20 |
| Violet, syn. | 250 |
| Civet tinc. | 20 |
| Violettyne MIP, Ferminich | 50 |
| Tobacarol | 50 |

## Chypre

I love this family of perfumes, and it has been within the past few years that these have gone out of style because of their strong and long-lasting scent. I believe they will come back in a few more years when we have a new generation of people that will appreciate once again the qualities of this great class of perfumes.

Following are two of my renditions of Gattefosse's chypre formulas, and of course, oakmoss and patchouli are of utmost importance. This first one reminds me of amber incense.

**Chypre #7 by Bobbie Kelley**

Bergamot ....................................................440
Sandalwood ....................................................80
Patchouli ....................................................10
Lemon ....................................................10
Cedar leaf ....................................................10
Rose ....................................................90
Jasmine ....................................................80
Clove ....................................................40
Vanillin ....................................................40
Coumarin ....................................................20
Labdanum absolute ....................................................20
Musk, synthetic ....................................................20
Oakmoss ....................................................30

**Chypre #8 by Bobbie Kelley**

Bergamot ....................................................220
Patchouli ....................................................10
Sandalwood ....................................................70
Ylang ....................................................60
Oakmoss ....................................................40
Clove ....................................................30
Jasmine, synthetic ....................................................50
Rose geranium CP ....................................................80
Cinalkex ....................................................20
Cassia ....................................................30
Coumarin ....................................................70
Heliotropin ....................................................100
Musk, synthetic ....................................................50

# CHAPTER 10
# Fantastical Perfume Journeys

I want perfumes to take me on a sensual journey. The following stories are the journeys that these particular scents that I have made have taken me on—whether real, imagined, or through dreams or fantasies. I believe that olfaction is directly related to all of our other senses, and we can both experience and create through them. I have named this chapter "fantastical" using the definition for the word *fantasy* (i.e., thought, creative visualization) as in the creative power of the imagination, which I believe is how all things are made, including perfumes. Thought has great power. I believe scents can make us project pictures, or what I call olfactory visions. How does this happen, you might ask?

There is a strange phenomenon called synesthesia, which has not been investigated much except by a man named Richard Cytowic, who wrote the book *The Man Who Tasted Shapes*, an excellent read regarding accounts of rare people who somehow get their senses crossed and experience extraordinary things such as tasting shapes. I believe if this could be figured out, then it would open a whole new dimension for the world of olfactory studies and the way humans perceive odors.

In regard to the reader asking if the following events have really taken place, I leave it up to them to decide.

### Flesh Parfum by Bobbie Kelley

I am blindfolded, and I am not sure where it is that I am. I am on a horse's back, sitting straddled in a leather saddle. There is someone very strong sitting behind me, and as the horse strides along, I can feel their body moving rhythmically next to mine. I cannot hear, but all of my other senses are heightened. I am neither tired nor hungry. There are bags of something hanging down on the sides of the saddle, and the acrid air sends its wafts up to my nose, but I say nothing. Somehow I know that I am supposed to not speak, just to sense. There seems to be no time. I have no fear. I am thirsty, and somehow my riding companion knows this and hands me a leather pouch to drink out of; and as soon as I hold the container to my lips, I know that it is arrack. I drink it as if I am drinking water. My body begins to feel as though it is melting into my master's body, and I have succumbed to it as the horse moves along as if there is no morrow. My master moves to the side to reach for the saddlebags. I hear some rustling and clanging against the warm leather. The smell coming from the bag becomes even more heightened as my master slowly rips my top covering, baring my breasts for the world to see. Suddenly I become aware and conscious of my breathing becoming harder and faster; my heart pumps its blood through my coursing veins, and I can hear my heartbeat in my head. My master pours a hot liquid onto the back of my neck, and as it runs down my spine, so does my mind. My mind runs away, and I am only left with my soul. The horse senses the excitement and starts to gallop forward. My master reaches his strong hands around to my breasts and starts to caress them with the hot liquid as he pulls on the reins. The smell of the hot liquid is strong and odiferous and, as the wind shifts, it mixes with the smell of my master, and I notice myself salivating like a hungry animal. I try to mutter sounds, but I am unable to as my master starts to kiss and bite the back of my neck. I can feel the horse's vibrations as it snorts. I do not fight my master. I am enslaved into my captivity. There is no escape for me. I am lost in a sea without water, without milk or honey.

### Nebulous Parfum by Bobbie Kelley

I am standing naked in a waist-high, sandy-bottomed pond in Eden's Garden. It is surrounded by huge fig trees, and below are gigantic pink peonies. There are blue lotus blossoms and mint floating all around me in the crystal-clear water. There are koi fish that swim past me, sensually rubbing themselves against the water hyacinths and around and in between my legs. My hair is long, and the strands look like corn silk in the water as the koi swish through them as if they are making their way through a labyrinth. Lovebirds sing their songs on the branches of the figs, cooing and wooing each other as if it is the first day of spring. The air is laced with sake. I smell the last of the winter's snow as it drips off the mountain. Jacque Cavalier approaches me from behind and whispers something softly in my receptive ears. I can't hear him, so I beckon to him to move in closer. He gently moves my hair aside and continues his murmurs. I move my hair for him to wash my back. I close my eyes and lean my head back and let him drip the pond waters over my forehead, and he does this with such precision. When I open my eyes, Issey Miyake is sitting on a bench in the distance, watching as I turn to wash Jacque. Issey has in his hand a glass of umbushi plum liqueur, and as he takes it to his moist lips, I see a glisten in his eyes that can only be described as sparkling tiger's eye growing out of a rock. I turn back to Jacque and pierce him with my eyes as if to climb into his spirit. Jacque and I become so light it is as if we are levitating out of the water. I become so lost into him that we become one and the same. His thoughts are my own, and mine his. There is no turning back for us. We now speak only with our minds. I tell him that I have been waiting for him for so long, but he already knows. Time has stopped, and we can no longer hear the sounds in the distance. Togetherness has engulfed us.

***Water Lily***

### Playful Parfum by Bobbie Kelley

I am in the forest of Poli Poli in the spring of 1989. "Do you want to do a hit of acid?" the doctor asked.
"Sure, why not," I answered. "Just allow me to meander through the field of clover and violets first, and I would like to stop and make us some flower necklaces for our journey through time and space."
"No problem, take your time," replied the doctor.

Traipsing along the non-path, I stumbled upon a small, little pond. It was clear and beautiful. I decided to put down my backpack and stick my face into the water and see what was up in the aquatic world these days. "Do you mind?" I asked the doctor.

"No, not at all, let me know about how that goes."

"I will be back in a couple of minutes," I said. I stuck my face in the water, and much to my surprise, amoeba were dancing and prancing and frolicking around, playing hopscotch in the water like little children playing in the streets.

"What are you guys doing?" I asked.

"Oh, we are playing around," they replied.

I then asked, "Am I going to get sick talking to you?"

They broke out into thunderous laughter. "Are you serious?" they asked.

Right at that second, I knew that I needed to come up for air and said, "Wait a minute, I will be right back. I have to go up for air." So I brought my head out of the water, took another deep breath, and plunged my face back into the water, anxious to talk to them again.

They then said, "That is the trouble with you humans. You always blame your illnesses on us, and we have tried to tell you before that you are the cause of your own misgivings. Now go and tell the rest of the world what we have said."

"I can't do that," I snapped. "People will think that I am crazy."

"Well, so what," they remarked. "That is not our problem. Now go, but do come back and visit us again sometime."

I came up for a breath and thought about what had just transpired. Could that really have happened? Did I really talk to amoeba? How can this be? How am I going to tell my friends that I conversed with amoeba? Oh, how bizarre! I had better keep this to myself.

***Water Lily***

## Butterfly Parfum by Bobbie Kelley

"What? This garden is for me? Oh, it is our birthday? That's right, I almost forgot that you and I were born on the same day. Oh, Rumi, my beloved, recite to me your poetry."
"Yes, my gentle butterfly. Come lie in my arms, and I will speak to thee."

Because of your love
I have lost my sobriety
I am intoxicated by the madness of love

In this fog
I have become a stranger to myself
I'm so drunk
I've lost my way to my home

In the garden
I see only your face from trees and blossoms
I inhale only your fragrance

Drunk with the ecstasy of love
I can no longer tell the difference between drunkard and drink between Lover and Beloved

"Oh, more, Rumi, don't stop, I beg you."

From the beginning of life
I have been looking for your face but today I have seen it

Today I have seen the charm, the beauty, the unfathomable grace of the face that I was looking for

Today I have found you and those who laughed and scorned me yesterday are sorry that they were not looking as I did

I am bewildered by the magnificence of your beauty and wish to see you with a hundred eyes

My heart has burned with passion and has searched forever for this wondrous beauty that I now behold

I am ashamed to call this love human and afraid of God to call it divine

Your fragrant breath like the morning breeze has come to the stillness of the garden
you have breathed new life into me
I have become your sunshine and also your shadow

My soul is screaming in ecstasy every fiber of my being is in love with you

Your effulgence has lit a fire in my heart and you have made radiant for me the earth and sky

My arrow of love has arrived at the target
I am in the house of mercy and my heart is a place of prayer

## Celestial Bodies Parfum by Bobbie Kelley

I am flying through the air. The atmosphere is pure white. I move faster and faster through the ether. It appears to be infinite. I am neither hot nor cold. Where am I? I asked myself silently. There is a feeling deep in myself that I already know the answer. I look at my sides, and I see beautiful snow-white wings gliding gently through the heavens. I have no fear. I receive a message from a lovely, feminine, angelic voice that tells me to turn around and go back to where I was, that it is not yet my time. I obey. As I am flying back toward the earth, I choose to go to the largest continent that I see. I am excited to land where there are no people, just animals. The land is vast and beautiful and covered with snow-white lilies. I hover over a sleeping lion. He awakens and allows me to stroke his majestic long mane. He does not seem to mind that I am levitating over him, and it seems that it is completely natural to him. I hug him and tell him telepathically that I love him. I receive the same message from him. I tell him that I have to go, but I will see him again someday. He understands.

***Easter Lilies***

## Seduce Parfum by Bobbie Kelley

I am in Cairo, Egypt, at the turn of the twenty-first century with my girlfriend. We are staying at the beautiful Marriott hotel that used to be a palace for a Jordanian prince. We get invited to a party on the top floor, and we gladly accept the invitation. Stepping out of the elevator, the floor shakes so hard that we think it is an earthquake, but we soon realize it is just the music. Loudly it plays, and then even louder as we move in closer to the party. There are at least one thousand Arab men singing and shouting at the top of their lungs. My friend and I are a little overwhelmed, but excited to be there and to witness this masculine energy. As we make our way through the smoke-filled room, the men catch sight of us. Several of them start wrapping a scarf around my hips and toss me upon a rather large tabletop to belly dance for them. As I start to move my hips, they go wild, and the crowd gets even louder. Then, as if that was not good enough, they decide that they want me onstage with the band, so they grab me and carry me swiftly to the stage to perform in front of everyone. My heart beats fast, and I cannot breathe as I look out onto the crowd. I then see my girlfriend up on top of the table that I was just moved from; she is laughing and dancing her heart out. I start to dance because I don't know what else to do. The men scream so loud that it is almost deafening. There are huge handheld cymbals banging next to my hips and at least twenty violins playing into my ears. It is truly one of the most exciting adventures my girlfriend and I have been on. I will never forget it.

## Nectar Parfum by Bobbie Kelley

"Kiss her," I heard Glen say.
"What?" I asked.
"Kiss her," my friend Glen said.
"Right here in front of all of the men?" I asked.
"Yes, especially them," Glen said again. "They will love it," he added.
"But we are at work, and this is a research center," I replied.
"So what, it is New Orleans. You can do anything you want here," Glen retorted. "And I want you to do it right on this patient's bed here," demanded Glen as he pointed his finger to the exact spot that he wanted it done.
I kept looking back at my girlfriend Giselle, and then to my friend Glen. Giselle kept smiling, looking at me to see what I was going to do. "But, Glen, you are gay, and you want to see me make out with Giselle in front of you?" I asked.
"Yes, I know, I know, but I think it is so sensual and sexy," Glen replied.
"Will it make you happy, Glen?" I asked.
"Yes," he responded.
I leaned forward slowly toward my girlfriend Giselle and looked into her obsidian eyes. She was looking straight into my eyes but, this time, way more seriously than before. The smile had left her porcelain face. Glen moved in closer to get a better view. As if in slow motion, I sensually touched Giselle's tender lips with mine. Her breath was perfumed. She was eager to receive, and then opened her mouth as if to accept food. Then the kiss became passionate as Giselle moaned loudly as her tongue moved forward. Right at that moment, the phone rang and broke our sensual focus. "I have to get back to work," I said to Giselle as I tried to catch my breath. "Meet me at Tiffany's bar tonight after work."
Giselle said nothing and turned and winked as she fled out of the door.
I turned and looked at Glen, who had just hung up the phone. "Are you happy now, Glen?" I asked.
"Very," replied Glen. "I owe you one," he said with a big grin on his face.

## Luscious Parfum by Bobbie Kelley

Istanbul, Turkey. It was a beautiful night along the Mediterranean, and it seemed that it would never end. I could not sleep, so I decided to take a walk along the lovely and beautifully manicured pathway made just for tourists. Night-blooming jasmine filled the air. A gorgeous young man followed not far behind. I was a little nervous, but at the same time excited at the way that he stalked me. I sensed he was harmless and did not smell any danger. Like a cat, I moved through the curling paths, intentionally making last-minute decisions as it forked. I decided to leave the path and go into the woods to see if I would be followed. There he was right behind me. I stood by a tree and turned and looked at him. He stopped to see what I would do next. He was striking, with large brown eyes, with framed glasses to match. I lifted my skirt a little, showing part of my legs. He moved forward slowly, not knowing quite what to do. He appeared to be somewhat younger than me. I motioned to him to step forward. As he advanced, I lifted my blouse, exposing my breasts. He ran to me and started kissing them ever so gently. I made him take off his shirt and bare himself to me. I ran my hands over his well-formed chest and then moved them down to his abdomen . . .

## Parfum Frangipani designed by Bobbie Kelley

I am at a wedding, and it is springtime. I am in a beautiful garden in a valley graced with the most lovely plumeria trees of all colors—pink, yellow, purple, red, orange, blue, green, lavender, and pure white. Scattered in the trees are bouquets of lush pink peonies and sweetly scented pink and lavender bromeliads. There are topiaries of all shapes and sizes cut from mock orange trees, and underneath them are huge sprays of different-colored hydrangeas. There are urns full of branches of flourishing cherry blossoms, some reaching ten feet high and spilling over the sides like an umbrella. Surrounding the urns are moats of clear water filled with calla lilies, water hyacinth, blue lotus, and golden koi fish. To my right, I see an infinite orchard of stargazer lilies, and to my left, an orchard of tuberose. There are damask rose petals scattered over the rich and fertile earth beneath my feet. There is a fine-looking mist that covers the valley like a sheer veil of virginal white spun silk. In front of me are more plumeria trees covered like a blanket with passionflower vines intertwined with stephanotis vines and crossing like shoelaces. At the ends of the branches, there are baskets dripping with fuchsia blossoms. Between the trees are trellises full of honeysuckle vine and forsythia reaching up from below as if to kiss the blossoms. The bride's bouquet is full of luxuriant gardenia, shy pink peonies, and showy pink cymbidium orchids embellished with pink lily of the valley. There seems to be no end to this day as I hear violins playing in the distance and answering to harps as they play back and forth in conversation . . .

*Fushia*

*Whirling Dervish Fushia*

*Passion Flower*

*Hydrangea*

*Scented Bromeliad (Tillandsia lindenii)*

*Plumeria*

*Double flowering cherry blossoms*

## Night-Blooming Cereus Parfum by Bobbie Kelley

"Hello," I answered my phone.

"Is this Bobbie?," I heard someone say on the other end of the phone.

"Yes it is," I answered.

"Your third test result is still class five cancer," she said in a firm voice.

"I am sorry," she blurted out before I could even say anything.

I hung up the phone without saying good-bye. I wondered and was asking myself as I walked into my room to pack my things, "Is this why the doctors keep telling me that I can't have children? Oh well, it doesn't matter." I reached for my backpack. As I drove through the day, I kept wondering if I should call my school counselor and tell her that I wouldn't be back at school Monday. I decided by the time I had reached my destination that I would not, nor would I call anyone and tell them the dreadful news.

I reached the middle of the desert by evening. The sky was vast and the air acrid and calm. The earth was covered in sand like a barren wasteland, but to my right, I had noticed a patch of night-blooming cereus stems, and beneath them, blanketed with peyote cactus, I lay down on the warm sand. There were a billion or more stars in the sky. I could hear what I thought were coyotes in the distance howling like they were a lonely gang of unloved and misunderstood beasts. I had closed my eyes for an unknown length of time and, upon opening them, was greeted with a huge black wolf hovering above my face. I lay very still while she sniffed me from my face down to my feet, and I was very unnerved to be at her mercy.

After a while, she came and sat by my side; and after a period of what seemed like eternity, she then spoke. "I am going to have to turn you inside out and help heal you, but you are going to have to agree and allow me," she said.

"But I came here . . ."

"Silence!" she shouted. "From this moment on, you are a wolf, and you will speak in wolf language," she firmly said.

I growled at her in agreement. By this time, nighttime had fallen. She raised her head into the wind and called for her clan to join us. I became acutely aware both by sight and scent of approaching wolves, and majestically, they filed one by one to my side and each one with a night-blooming cereus flower in their mouths. They each took turns and dropped them in the sand all around me, and as each one did this, they would growl at me, and I in turn growled back to acknowledge them and their presence.

Mother Wolf then put her jowls to my lower abdomen and started snarling as if she was warding off demons, and as she did this, the pack of wolves followed her lead, and all grumbled in unison. Their faces grimaced and contorted, and their canines were shining in the moonlight like thorny pearls glistening in water as they drooled and slobbered over my helpless body. Then Mother Wolf said in wolf language, and I understood, "Growl with us at this monstrous beast, and let it be aware that it can never enter you again!"

As I snarled, I suddenly became aware of beautiful lights flashing about me from the night-blooming cereus flowers as though they were beaming healing light in my abdomen, replacing the horrific creature that was to leave my body.

All photographs were taken by the Author.
The images do not necessarily reflect the subject matter.

# Reference Material and Suggested Reading


*Formulary of Perfumes and Cosmetics*. Rene-Maurice Gattefosse. Chemical Publishing, 1959.

*Gattefosse's Aromatherapy*. Rene-Maurice Gattefosse, from the original French text *Aromatherapie: Les Huiles essentiells hormones vegetales*. C. W. Daniel, 1937.

*Perfumes, Cosmetics and Soaps*. Volume 1. W. A. Poucher, Seventh Edition. Chapmen and Hall, 1974.

*Perfumes, Cosmetics and Soaps*. Volume 2. W. A. Poucher, Ninth Edition. Chapmen and Hall, 1993.

*Handbook of Perfumes and Flavors*, Dr. Olindo Secondini. Chemical Publishing, 1990.

*Perfumery Practice and Principles*. Robert R. Calkin and J. Stephan Jellinek. Wiley-Interscience Publication, 1994.

*Perfume and Flavor Materials of Natural Origin*. Steffen Arctander. Self-published by the author, 1961.

*The Manufacture of Perfumes*. John Snively PhD. Charles W. Smith, 1877.

*Perfumes & Their Preparation*. George W. Askinson (Dr. Chemistry). Merchant Books, reprinted 2006.

*Perfumery and Flavoring Synthetics*. Paul Z. Bedoukian Ch. E., PhD, Allured Publishing Corp., 1986.

*Perfumery & Flavoring Materials*. 50 Years of Annual Review Articles 1945–1994, Paul Z. Bedoukian Ch. E., Ph.D., Allured Publishing Corp.

*Flower Oils and Floral Compounds in Perfumery*. Danute Pajaujis Anonis (perfumer and flavorist). Allured Publishing Corp., 1993.

*Praktikum des Modernen Parfuemeurs*. P. Jellinek. Urban & Schwarzenberg, 1949.

*Perfumery: Techniques in Evolution*. Arcadi Boix Camps. Allured Publishing Corp., 2000.

*The Scent of Orchids*. Roman Kaiser. Elsevier, 1993.

*Das Komponieren in der Parfuemerie*. O. Gerhardt. Leipzig: Verlagsgesellschaft, 1931.

*The Chemistry of Fragrances*. David Pybus and Charles S. Sell. Royal Society of Chemistry, 1999.

*Fragrance Applications: A Survival Guide*. Stephen J. Herman. Allured Publishing Corp., 2002.

*Speciallty Aroma Chemicals in Flavors and Fragrances*. Dr. Yunus Shaikh. Allured Publishing Corp., 2002.

*Preservatives for Cosmetics*. David C. Steinberg. Allured Publishing Corp., 2002.

*The Art of Perfumery and Method of Obtaining the Odors of Plants*. G. W. Septimus Piesse. Lindsay and Blakiston, 1857.

*Formulaire de la Parfumerie*. R. Cerbelaud. Paris: Editions Opera, 1951.

*A Practical Guide for the Perfumer*. H. Dussauce (professor of chemistry). Henry Carey Baird Industrial Publisher, 1868.

*Allured Flavor and Fragrance Materials*. Allured Publishing Corp., 2006.

*The Compleat Distiller*. Nixon & McCaw, second edition. Amphora Society, 2004.

*Odoratus Sexualis*. Dr. Iwan Bloch, 1933.

*The Man Who Tasted Shapes*. Richard Cytowic. New York: Jeremy Tarcher, 1993.

*Essence and Alchemy*. Mandy Aftel. North Point Press, 2001.

*Flower Confidential*. Amy Stewart. Workman Publishing, 2007.

*Perfume*. Nigel Groom. Running Press, 1999.

*Fabulous Fragrances*. Countess Jan Moran. Crescent House, 2000.

*Conversations with Mummies*. Rosalie David and Rick Archbold. William Morrow, 2000.

*Encyclopedia of Fruit Trees and Edible Flowering Plants*. Alfred G. Bircher and Warda H. Bircher Ph.D., American University in Cairo Press, 2000.

*Napoleon's Buttons: 17 Molecules That Changed History*. Penny Le Couteur and Jay Burreson, Penguin Group, 2003.

*The Web That Has No Weaver*. Ted J. Kaptchuk. Congdon & Weed, 1983.

*The Tao of Sexology*. Dr. Stephen T. Chang. Tao Publishing, 1986.

*The Love Poems of Rumi*. Deepak Chopra M.D. Harmony Books, 1998.

*The Life and Teachings of the Masters of the Far East*. Baird T. Spalding, six-volume set, Devorss, 1924.

*Merriam Webster's Collegiate Dictionary*. Tenth Edition. Merriam-Webster, 1997.

*The Emperor of Scent*. Chandler Burr. Random House, 2003.

*Perfume:* The Story of a Murderer, Patrick Suskind, Random House Inc., 1986

*Jitterbug Perfume,* Tom Robbins, Bantam Books, 1984

# INDEX

**D**



**E**



**F**

Made in the USA
Las Vegas, NV
18 October 2023